Sourcing to Support
the Green Initiative

Sourcing to Support the Green Initiative

Lisa Ellram, PhD, CPM, CMA
Wendy Tate, PhD

business**expert**
Press

First published in 2013 by
Business Expert Press, LLC
222 East 46th Street, New York, NY 10017
www.businessexpertpress.com

ISBN-13: 978-1-60649-600-8 (paperback)
ISBN-13: 978-1-60649-601-5 (e-book)

Business Expert Press Supply and Operations Management collection

Collection ISSN: 2156-8189 (print)
Collection ISSN: 2156-8200 (electronic)

Cover and interior design by Exeter Premedia Services Private Ltd.
Chennai, India

First edition: 2013

10 9 8 7 6 5 4 3 2 1

Printed in the United States of America.

To Jeff and Celeste Siferd. Without your unwavering support, sense of humor, and willingness to listen to my stories about global supply chains and the environment, this would not have been nearly as fun.

Lisa M. Ellram

To my daughters, Whitney and Tayler, thanks for all the support, love and interest in supply chain and environmental issues. The discussions and questions we had helped make this book more interesting and relevant.

Wendy L. Tate

To all of the practitioners who have shared their time and experience with us, and are striving to make a difference in their organization's environmental footprint. Thank you.

Abstract

The "green" or environmental sustainability movement has taken hold throughout the world. Its staying power is confirmed by the fact that environmental emphasis in organizations did not decrease during the recent global recession, but rather increased. Organizations see that greater environmental emphasis in their product and service offerings not only helps their image, but also may reduce inefficiency, waste, and corresponding cost. However, since most organizations rely heavily on their supply base for providing components, materials, and services that become part of their final products, organizations must revisit their sourcing processes and choices to have a real impact on the environment.

Sourcing to support an organization's green initiatives begins with the idea conception stage of new product development. Just as organizations consider the features that their products will provide to customers, they should also consider the environmental footprint of their products and services and design them accordingly. The idea conception and design generation may engage key suppliers to develop the right environmental footprint. The buying organization must make it clear to potential suppliers what is expected of them in terms of their environmental standards and what the buying organization is willing to assist with, if anything.

In some cases, the buying organization may be dealing with suppliers who have advanced environmental capabilities. If that is the case, the buying organization can potentially learn much from its suppliers. Suppliers who have many customers in certain industries, such as electronics or food, have been held accountable to stringent industry standards for years, and may know more about improving a product's environmental footprint than the buyer. Buying firms should leverage this knowledge.

This book begins with an introduction to the idea of sustainability and to the concept of what it means to source to support an organization's green initiatives, and why this is important from a holistic, lifecycle perspective. The initial chapters will also provide a point of view of how green sourcing fits into the organization's entire portfolio of sustainability initiatives. Examples of industries and associations that lead the way in green sourcing will be presented. In addition, best practices in green sourcing will be discussed. The book also provides a perspective on how organizations can encourage and support their suppliers in pursuing green initiatives, and what types of initiatives provide a good starting point.

The final chapters of the book will present more specific examples of the way in which various industries and groups view green environmental purchasing holistically. This includes a view into how the organization's transportation decisions and choices can support green sourcing and the overall goal of the organization to reduce its environmental footprint.

Keywords

green purchasing, sustainable purchasing, sourcing for the environment, green suppliers, environmental transportation.

Contents

List of Illustrations

Tables

Figures

Abbreviations and Acronyms

BSR	Business for Social Responsibility
c2c	Cradle to cradle
CO_2	Carbon dioxide or greenhouse gas
DfE	Design for environment
EICC	Electronics Industry Citizenship Coalition
EP	Environmental purchasing
EPA	Environmental Protection Agency, a unit of the federal government
EPEAT	Electronic product assessment tool
EPP	Environmentally preferable purchasing: a set of guidelines prescribing preferred purchasing practices for federal government agencies
EPR	Extended Producer Responsibility
GHG	Greenhouse gas, also known as CO_2 or carbon dioxide
IPR	Individual producer responsibility
NASPO	National Association of State Procurement Officials
NGO	Nongovernmental organization
NO	Nitrogen oxide
PM	Particulate matter
RoHS	Restriction of Hazardous Substance
Scope 1 Emissions	Emissions from the company's owned operations
Scope 2 Emissions	Indirect GHG emissions resulting from the generation of electricity, heating and cooling, or steam generated off site but purchased by the entity
Scope 3 Emissions	Include indirect GHG emissions from sources not owned or directly controlled by the entity but related to the entity's activities. This includes, but is

	not limited to, employee travel and commuting, contracted solid waste disposal, and contracted wastewater treatment
SER	Social and environmental responsibility
SMRS	Sustainability Measurement & Reporting System, used by the Sustainability Consortium
USCAR	United States Council for Automotive Research
VRP	Vehicle Recycling Partnership
WEEE	Waste Electrical and Electronic Equipment Directive

CHAPTER 1

Why Should We Care?

Introduction

In this chapter, some background information is provided to support the idea that green sourcing is an area of growing concern and importance, and that a firm's buying practices can have an impact on the environment. The issue of green sourcing is becoming more urgent as customer pressure for environmentally sound products grows. The evidence continues to mount that solid waste and pollution are growing at a faster rate than the planet and current technology can safely dispose of, or regenerate, needed inputs.

To address the issue of the growing scarcity of resources and increasing pollution, one approach is to generate less waste. Another approach is to use materials that can be reused, remanufactured, and recycled. Yet a third approach, which few businesses currently advocate, is to consume less as a society. Because this book is written for business professionals from all fields, the focus will be on the first two methods of reducing waste, while providing some insight into the third approach from a business perspective.

Growing Interest in the Environment

Sustainability is a significant and growing issue for many organizations today. Sustainability is described as "creating and maintaining the conditions under which humans can live in productive harmony with the natural environment while fulfilling the social, economic, and other requirements of present and future generations."[1] The focus of this book is specifically on the businesses' green initiative, or improving the environmental pillar of sustainability, while remaining economically viable. The specific context is what the purchasing function can do to help businesses achieve their environmental goals.

Each day, most major publications have news stories about the natural environment—its degradation, companies that are trying to minimize environmental impact, and NGOs that are working to make a difference. Facts such as the average U.S. consumer generates about 4.43 pounds of solid waste daily, of which 1.51 pounds is recycled or composted, are routinely heard in many different forums.[2] The fact of the matter is that while household waste is high, of even greater concern is industrial waste.

The per capita solid waste in European Union (EU) countries is about 75% of that in the United States.[3] Amazingly, businesses generate significantly more waste and pollution than consumers. For example, the latest EU statistics (Figure 1.1) indicate that households generate about 8.5% of solid waste, while construction, mining, and manufacturing generate 32.9%, 27.8%, and 13.1%, respectively.[4] While much of what these industries do is ultimately designed to benefit households, consumers currently have very little say in the manner in which these industries conduct business.

Businesses clearly have a significant impact on the amount of waste generated in a society. Therefore, they can play a major role in reducing the waste and other forms of pollution that they generate, and in the waste that households or consumers generate while consuming the products and

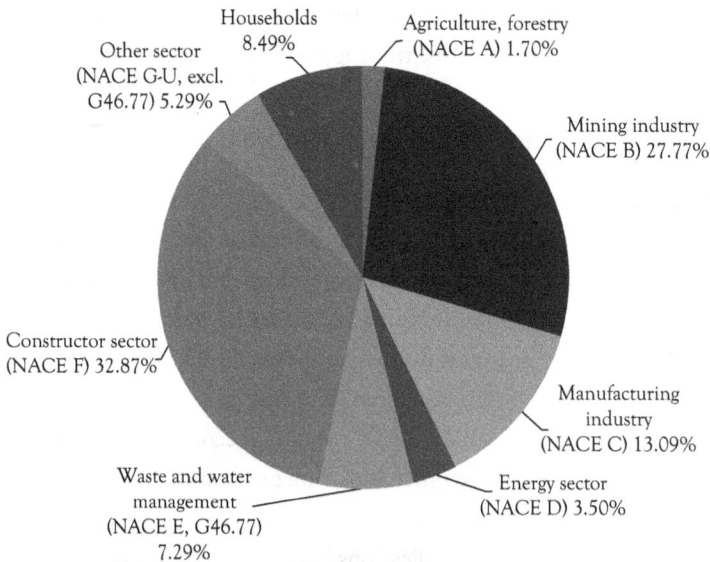

Figure 1.1. EU statistics on waste generation.

services that businesses create. Increasingly, both business and industrial consumers are demanding more environmentally friendly products and services. The staying power of environmental concern as a public issue is confirmed by that fact that the environmental emphasis of businesses did not decrease during the recent global recession, but rather continued to increase. A recent survey of Chief Supply Chain Officers (CSCOs) indicated that 71% agree that a major driver for their company's environmental sustainability efforts is to "Create a positive customer image and enhance brand equity."[5] Organizations see that greater environmental emphasis in their product and service offerings not only help their image, but also may reduce inefficiency, decrease waste, and positively influence the corresponding cost. But what does that really mean and how can businesses create an environmental strategy that meets these types of goals? And what is the optimal path for an organization to take to better meet environmental customer demands?

Potential Impact of the Purchasing Function

The role that the purchasing function (also known as supply management or procurement) plays within the organization in helping to support and promote an organization's green objectives is explored in this book. The role of purchasing is extremely influential in both manufacturing and service organizations. Because so few companies are vertically integrated, most of what companies actually make and sell is made up of parts and services that they purchase from other organizations in the supply chain. For example, based on the recent U.S. Survey of Manufactures data, materials costs average about 59.3% of the value of sales revenue for U.S. manufacturing firms.[6] For U.S. retailers, purchases average 64.5% of the total value of their sales revenue.[7] Clearly, since most organizations rely heavily on the suppliers in their supply base for providing components, materials, goods, and services that become part of their final products, organizations need to revisit their sourcing processes and choices to have a real and positive impact on environmental outcomes. Just as businesses generate much waste and pollution in serving consumers' needs, suppliers generate most of the waste in the supply chain, on behalf of their customers.

It is not unusual for an organization's suppliers and supply chain to have a greater cumulative impact on the environmental footprint of the organization's product or service than the organization itself. It was this realization that made Walmart take an extra hard look at its suppliers' environmental practices and then start an initiative to help suppliers improve their environmental impact. In late 2008, Walmart announced its global responsible sourcing initiative at the China Summit. It was during this time that it announced new goals for greater environmental (and social) compliance, transparency, and accountability.[8] This will be discussed later in the book, in Chapter 5. But not all suppliers and materials are equal in terms of their impact. Thus, the purchasing function can play a very influential role in the organization's environmental footprint through the supplier selection, management, and retention processes.

Before the supplier selection process even begins, people in the organization are making decisions regarding the specifications of products and services, the materials and processes to be used, and the price the item can ultimately be sold for. These decisions have a significant impact on the viable suppliers that purchasing can consider. As a result, the design decision also has an indirect, and possibly a direct impact on the pool of viable suppliers. This is all part of the product lifecycle.

Thus, for maximum impact, sourcing to support an organization's green initiatives begins in the idea conception stage of new product development. Just as organizations consider the features that their products will provide to customers, they should also consider the environmental footprint of their products and services and design them accordingly. The idea conception and design generation may engage key suppliers and potentially key customers to develop products and services with the impact on its environmental footprint as a key decision-making criterion. The buying organization must clearly specify to potential suppliers what is expected of them in terms of their environmental standards and what the buying organization is willing to assist with, if anything. These environmental expectations are also part of the supplier scorecard or performance measurement system.

In some cases, the buying organization may be dealing with suppliers who have advanced environmental capabilities. If that is the case, the buying organization can potentially learn much from its suppliers. Suppliers who have many customers in certain industries, such as electronics or food, have been held accountable to stringent industry standards for years, and may know more about improving a product's environmental footprint than the buyer. The buying firms should understand and also leverage this knowledge.

Environmental Footprint

A product's environmental footprint is the sum total of all of the impacts that the product has on the environment at every stage of the supply chain and in every phase of the product lifecycle.[9] It is a complex concept, and numerous organizations are working on its definition and measurement. Importantly, the definition and how it is measured is holistic and balances how a decision at one stage in the lifecycle can affect the other stages. The concept of environmental lifecycle is presented in more depth in Chapter 2.

Historically, if companies have measured their product's environmental footprints at all, it has been based on their own operations, often referred to as Scope 1 emissions. However, this is changing so that any mean expression of a product's or an organization's environmental footprint considers the inputs and outputs required over the entire lifecycle.

Breadth of the Scope of Footprint Emissions

Companies also define the scope of emissions that they measure differently. A commonly used approach for defining how much greenhouse gas (GHG) emission an entity is including in their reports is provided by the Environmental Protection Agency (EPA):

Scope 1. Emissions are direct GHG emissions from sources that are owned or controlled by the entity. Scope 1 can include emissions from fossil fuels burned on site, emissions from entity-owned or entity-leased vehicles, and other direct sources.

Scope 2. Emissions are indirect GHG emissions resulting from the generation of electricity, heating and cooling, or steam generated off site but purchased by the entity, and the transmission and distribution losses associated with some purchased utilities (e.g., chilled water, steam, and high temperature hot water).

Scope 3. Emissions include indirect GHG emissions from sources not owned or directly controlled by the entity but related to the entity's activities. Scope 3 GHG emission sources currently required for federal GHG reporting include transportation and distribution losses associated with purchased electricity, employee travel and commuting, contracted solid waste disposal, and contracted wastewater treatment. Additional sources that are currently optional under federal reporting requirements, but are significant, include GHG emissions from leased space, vendor supply chains, outsourced activities, and site remediation activities.[10]

While these definitions provided by the EPA refer only to GHG emissions, the definition of the environmental footprint above would have them include much more. These definitions are readily modified to apply to any type of waste generation. For example, the EU suggests capturing a host of outputs at every stage of the product lifecycle and product use. These factors include, but are not limited to, GHG, water usage and depletion, mineral depletion, acidification, radiation, particulate matter, and more.[11] Today, most companies only report on their Scope 1 emissions. However, that is changing as will be presented in future chapters. Honda is one of the few companies that has begun to report its Scope 3 emissions, the first "mobility" company to do so. This provides a very different and more holistic perspective that really imparts insights into the impact of a product over its lifecycle. For example, Honda reports that its Scope 1 GHG emission is 1.24 million tons, while its emission from customer use of products (part of Scope 3 emissions) is 195.88 million tons.[12] This can provide a company with an excellent perspective on where they should focus their efforts. In Honda's case, improving gas mileage of its cars and providing owners with information on how to maximize their gas mileages could have a huge potential impact.

Implications of Green Sourcing

The impetus for green sourcing goes beyond increasing customer goodwill and includes issues such as complying with regulations in all of the countries in which an organization conducts business, reducing the risk to the organization of supplier noncompliance, and maintaining and enhancing the organization's reputation.

From a regulatory standpoint, there are standards in place that dictate the way in which waste should be handled and the types of materials that can be used in products. These issues should be addressed in product design, supplier selection, and supplier audits. A firm can find that if its suppliers are noncompliant, the supplier may be fined, delaying deliveries and increasing costs. In addition, the firm's name can be associated with a supplier with a tarnished reputation, which can hurt the firm's reputation. Thus, supplier noncompliance can place the buying organization at risk. Increasingly, companies want to have visibility into their supplier's environmental performance. Twenty-five percent of CSCOs who responded to a recent survey indicated they have visibility into the environmental performance of their extended supply network, beyond the first tier, while an additional 28% indicated that they have visibility into the environmental performance of their first tier suppliers only.[13] This is a very positive trend for the natural environment.

As many of the companies in the above survey realize, organizations that purport to be interested in reducing their environmental footprint must look beyond their own four walls. Even highly vertically integrated organizations still buy things from suppliers and still ship things that they buy and sell. Because of the huge impact that suppliers have on the total environmental footprint of the supply chain for products that a company buys and sells, the impact of supplier practices and location cannot be overlooked if the organization wants to meaningfully affect the supply chain footprint.

Organization of the Book

This book begins with an introduction to the idea of sustainability and to the concept of what it means to source to support organization's green initiatives, and why this is important from a holistic, lifecycle perspective. The initial chapters will also provide a point of view of how green sourcing fits into the organization's entire portfolio of sustainability initiatives. Examples of industries and associations that lead the way in green sourcing will be presented. In addition, best practices in green sourcing will be discussed.

The book also provides a viewpoint on how organizations can encourage and support their suppliers in pursuing green initiatives, and what types of initiatives provide a good starting point. The final chapters of the book present a view on end-of-life product issues and how they relate to green sourcing; how the organization's transportation decisions and choices can support green sourcing; and the overall goal of the organization to reduce its environmental footprint.

Key Points

- Sustainability, which includes environmental, social, and economic concerns, is growing in importance and recognition globally.
- Levels of emissions and landfill pollution are expanding much more rapidly than the earth's ability to effectively eliminate their negative impact.
- This book focuses primarily on enhancing the environmental aspects of purchasing, while ensuring that the organization remains economically viable.
- The purchasing function can have a tremendous impact on the organization's environmental footprint through selection of materials and suppliers.
- Organizations often classify their waste and emissions as Scope 1 (internally generated through manufacturing), Scope 2 (Scope 1 plus owned transportation and commuting pollution), and Scope 3 (everything else, including product use and disposal).
- Scope 2 and Scope 3 pollution are receiving more attention as organizations attempt to understand and influence their true product or supply chain footprint.

CHAPTER 2

What Is Sourcing to Support the Green Initiative?

Introduction

This chapter begins with defining "sourcing for the green initiative." This definition has its foundation in environmental purchasing (EP). Environmental purchasing and each of its elements will be defined and discussed. The EP definition will then be adapted to encompass the role of purchasing and the supply chain in green sourcing initiatives. The lifecycle concept, also known as a cradle-to-cradle perspective, is included in this chapter to provide a better understanding of how environmental management should be viewed over the entire product lifecycle. The lifecycle perspective looks at products from design and initial extraction of materials through to consumer use, return, and ultimately disposal. The intent is to analyze the inputs and outputs of materials and energy, and the environmental impacts that are directly attributed to a product, process, or service.[1] Decision-makers then can make the right choices for reduced environmental impact.

A Continuum of Perspective Related to Sourcing for the Green Initiative

The environmental pyramid is often shown as a hierarchy of desirable possible behaviors associated with handling waste. The least desirable way to deal with waste is disposal in a landfill. Moving up the pyramid is the safe disposal of waste with energy recovery, which is preferred, but recycling is better still. Reuse is preferred to recycling, as it maximizes the value of the item in its current form. The pinnacle of the pyramid, resource reduction, or ultimately zero waste is the ideal goal. However, this is not always possible. The idea is to not buy things that you will not use, and to use the things that you buy in a way that minimizes waste and maximizes the lifetime usability of the item. There is the potential to reduce by penalizing for waste and giving incentives for reduced waste.[2] The elements of the pyramid are defined and examples are provided of how some exemplary organizations are putting these practices to work. The pyramid and the elements of the pyramid then lead directly to the definition of environmental purchasing and help to link the concepts of lifecycle management to more effectively source for the green initiative.

Dispose/Landfill. Physical elimination of material is the least preferable method to deal with waste. Often we don't give much thought to where the material and products that we use end their life. In the United States, landfills are closing at an increasingly rapid rate and yet the amount of trash generated is increasing.[3] Some countries have banned landfills classifying waste as "raw material."[4] The landfills actually inhibit and delay the degradation of materials because the waste is too tightly packed, inhibiting oxygen flow. Landfills also contribute to reduced quality of the land, air, and water. One of the biggest landfill problems is the amount of packaging used to safely move products from a supplier into the hands of the customer. In recognition of the problems associated with landfills, an increasing number of companies are trying to reduce waste and achieve the goal of zero waste to landfill. For example, Sunny Delight achieved its

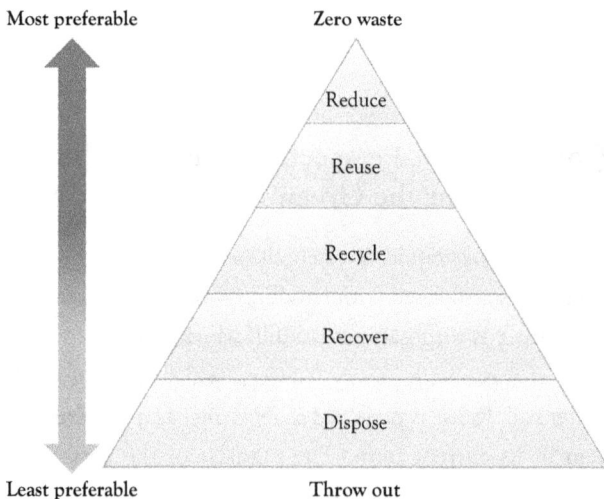

Figure 2.1. The environmental pyramid.

zero waste to landfill in manufacturing objectives three years ahead of schedule. It discovered that it could actually save money by not paying waste disposal and haulage fees.[5]

Recover. There is significant debate about waste recovery. Recovery generally refers to the following operations: material recovery (recycling), energy recovery (i.e., reuse of fuel), and biological recovery (composting). The debate is more about the terminology that comes from the specification of whether the waste that is recovered is a suitable replacement for nonwaste materials that would have to be used to achieve the same end benefit.[6] The issue is that the waste that is recovered helps to conserve natural resources. One way to reduce waste to landfill is energy recovery through incineration. However, this can create emissions issues. Thus, companies really committed to zero waste try to focus on managing waste higher up the pyramid, as in recycling and reuse versus trying to recover what they can but potentially have a negative environmental impact.

Recycle. In its most basic form, recycle means to put or pass through a cycle again. It is the process of reusing a product beyond its intended use or producing a new product from a material that would otherwise have been considered waste.[7] The idea behind recycling is to prevent disposal of potentially useful materials, reduce landfill space requirements, reduce the consumption of virgin raw materials, reduce energy usage, and reduce air pollution (from incineration) and water pollution (from landfilling) by decreasing the need for waste disposal. There is also the potential to decrease greenhouse gas emissions when compared to the utilization of virgin production. Sorting and separating items to be recycled is an important part of zero waste. Sprint's wireless recycling program helps prevent thousands of tons of wireless equipment from entering the waste stream. More than 90% of the devices are reused, and anything not reused is recycled.[8] Sprint also partners with the U.S. EPA's "plug in to e-cycling program."

Reuse. Reuse is preferred to recycle because it takes less energy and resources. Reuse is simply using an object or material again, either for its original purpose or for a similar purpose, without significantly altering the physical form of the object or material. Reuse prevents items from becoming waste. An example is bringing a reusable coffee mug to Starbucks for a refill instead of using a paper cup. Early in its campaign to reduce paper usage, Starbucks reported prevention of 550,000 pounds of paper waste because of this practice in one year.[9] Companies like Honda increasingly employ reusable shipping bins for repetitive processes within the supply chain, rather than one-use cardboard.[10] CHEP became an industry leader in providing reusable pallets, containers, and crates.[11]

Reduce. Waste reduction is defined as the minimization of waste at its source to reduce the quantity required to be treated or disposed. This is usually achieved through better product design and/or process management.[12] This is also a fundamental principle of lean production and manufacturing systems. Kraft Foods is continuing to focus on eliminating waste at its manufacturing plants. Since 2005 they have already seen a 50% reduction. Kraft Foods' strategy is simple: Generate less waste and find new uses for the waste it does produce. The company is getting results by changing its behavior, business practices, and culture.[13]

These different perspectives on waste exist on a pyramid. Reduction is the most preferable whereas allowing waste to enter a landfill is the least preferable. Environmental purchasing involves management of the waste stream from original extraction up to an including the disposal of the goods and materials. Understanding the implications of purchasing decisions on the environmental pyramid is essential to effective environmental purchasing. When a company decides to focus on reducing waste to landfill, it often has to change many of the items it is currently purchasing to equally effective but more environmentally friendly alternatives. The definition of environmental purchasing is provided below.

Defining Environmental Purchasing

Environmental purchasing (EP) for an individual firm is the set of purchasing policies formulated, actions taken, and relationships formed in response to concerns associated with the natural environment. These concerns relate to the acquisition of raw materials, including supplier selection, evaluation, and development; suppliers' operations; in-bound distribution; packaging; recycling; reuse; resource reduction; and final disposal of the firm's products.[14] This means that purchasers should:

1. Consider environmental impact in all supply and supplier decisions
2. Buy products that are made in environmentally preferable ways

3. Buy products that can be easily reused
4. Buy products that can be easily recycled
5. Buy products that are energy efficient
6. Educate others on how to use environmentally friendly products (energy efficient, recyclable, reusable) in ways that maximize their potential for recovery
7. Consider the environmental impact of purchased items over their entire lifecycle, from how the product is made, to its acquisition, use, and final disposition

Suppliers are beginning to see more requests for proposal (RFPs) that ask them to provide information about their green initiatives.[15] This is an indication that sourcing for the green initiative is beginning to take a more prominent position in organizational strategy. Green sourcing is an evolution of what is typically considered traditional strategic sourcing. It not only focuses on the more traditional waste management techniques defined above, but also can help organizations to drive cost reductions, including product content substitution, waste reduction, and lower usage. It can also turn waste products into sources of revenue and develop green "credentials" that will attract more customers.[16] In today's environment however, this definition does not encompass all of the efforts necessary to be effective in green sourcing initiatives. It also requires a focus on the closed-loop supply chain, returns management, and lifecycle perspective.

Environmentally Preferable Purchasing (EPP)

The Environmental Protection Agency (EPA) has a site for environmentally preferable purchasing that helps buyers consider the environmental impact of their purchases of items related to the categories listed in Table 2.1. While the EPP guidelines (available at their website) are prescriptive for federal government agencies, they are useful to and accessible by anyone.

In addition, there are many private and government-sponsored eco-labeling systems that a supply manager can use for guidance wherever they are in the world. Table 2.2 provides a listing of some of these. These resources are excellent when a buyer does not have the time or resources

Table 2.1. EPP Available Product Categories

• Building and construction
• Carpets
• Cleaning
• Electronics
• Electronic Product Environmental Assessment Tool (EPEAT)
• Fleets
• Food services
• Landscaping
• Meetings and conferences
• Office supplies
• Paper[17]

to invest in researching which options are the most environmentally friendly, when the buy is not a recurring buy, or when someone is new to the whole concept of EP.

Implementing Environmental Purchasing

Many state, local, and national governments throughout the world have taken a leadership role in implementing green, or environmental purchasing, as Table 2.2 suggests. Like any major process change, implementing EP requires a well thought-out process and plan. It requires clear communication with others in the organization regarding the goals, approaches, what will be measured, and how it will be measured. All of these issues will be presented in this book.

The way in which government agencies implement green purchasing may be different from the approaches followed by manufacturers, because the government generally does not produce any physical items itself. Yet the same underlying philosophy for implementing EP applies. The first implementation approach described is based on guidelines from the National Association of State Procurement Officials (NASPO). Generally, the drive to implement EP will come from higher levels in the organization. However, if it does not, SM can make the case itself to implement EP.

Table 2.2. International EPP Programs

Blue Angel (Germany)	The Blue Angel is Germany's eco-label. Created in 1977, Blue Angel is one of the oldest eco-labeling programs in the world.
Eco Mark (Japan)	The Japan Environment Association develops environmental standards and permits products to bear the Eco Mark symbol.
Environmental Choice (New Zealand)	The New Zealand Ecolabeling Trust is a voluntary, multiple specifications-based environmental labeling program, initiated and endorsed by the New Zealand government.
Environmental Choice Program (Canada)	The Environmental Choice Program (ECP) is Environment Canada's eco-label program.
Environmental Label Award (Croatia)	The Environmental Label is awarded by Croatia's Ministry of Environmental Protection for products certified as meeting specific environmental performance criteria.
European Commission, Green Public Procurement	Policies, resources, and guidance from the European Commission to increase the level of green public procurement in their member states.
European Union (EU) Eco-Label	The EU Eco-Label is used throughout the European Union as well as in Norway, Liechtenstein, and Iceland.
Global Ecolabelling Network (GEN)	GEN is a nonprofit association of third-party, environmental performance labeling organizations founded in 1994 to improve, promote, and develop the eco-labeling of products and services.
Good Environmental Choice (Australia)	Good Environmental Choice is Australia's lifecycle environmental performance ecolabel.
Hong Kong Green Label Scheme (HKGLS)	HKGLS is an independent, not-for-profit and voluntary scheme for the certification of environmentally preferable products launched in 2000 by the Green Council and the Hong Kong Productivity Council.
Green Mark (Taiwan)	Green Mark is Taiwan's program to guide the consumers in purchasing green products and to encourage manufacturers to design and supply environmental benign products.
Green Purchasing Network (Japan)	The Green Purchasing Network (GPN) was established in February 1996 to promote green purchasing among consumers, companies, and governmental organizations in Japan.

(Continued)

Table 2.2. International EPP Programs (Continued)

Korea Eco-Label	The Korea Eco-Labeling Program certifies qualifying eco-products for excellent quality and performance, as well as general environmental friendliness during the entire production process.
National Programme for Labeling Environmentally Friendly Products (Czech Republic)	The Czech Ecolabeling Agency manages the Czech National Ecolabeling Programme, which labels "environmentally friendly" products.
NF Environment Mark (France)	The NF Environment mark is official French ecological certification.
Organization for Economic Co-Operation and Development (OECD)	The OECD Environment Directorate provides governments with the analytical basis to develop environmental policies that are effective and economically efficient.
Green Label (Singapore)	The Singapore Green Labeling Scheme was launched in 1992 by the Ministry of the Environment, and awards the Green Label to products meeting specific environmental criteria.
Swan Eco-Label (Nordic Countries)	The Swan is the official Nordic eco-label introduced by the Nordic Council of Ministers.
TCO Development (Sweden)	TCO Development provides certification and environmental labeling of office equipment designed to improve the work environment and the external environment.
TerraChoice (Canada)	TerraChoice is a science-based environmental marketing firm.
Thai Green Label Scheme (Thailand)	The Thai Green Label Scheme was launched in 1994 by the Thailand Environment Institute in association with the Ministry of Industry, and awards the Green Label to products meeting specific environmental criteria.

Potential Benefits of EP

Some of the benefits that can help justify EP are detailed in Table 2.3. It is fairly straightforward to see that many of the environmental benefits could be easily translated into direct goals of the government. Alone, the environmental benefits may not be enough to justify the changes to private enterprises. This is especially true given the belief that many hold that

Table 2.3. Benefits of Environmental Purchasing[18]

Environmental benefits	• Manufactured using fewer toxic ingredients, minimizing hazardous impacts during manufacturing (e.g., water/air pollution), disposal, the risk to workers handling and using the products. • Improved energy efficiency lowers emissions of sulfur dioxide (which causes acid rain) and carbon dioxide (a primary greenhouse gas). • Manufacture of postconsumer recycled content instead of pure virgin products reduces the need to extract raw materials, in general, uses less energy and water, and reduces the amount of waste in our overstressed landfills. • Reduced landfill through reducing packaging weight and size. • Designed to reuse and/or reduce the amount of water needed to perform certain tasks, thereby conserving this very valuable resource. Such products and technologies involve plumbing devices, cooling systems, appliances, water treatment technologies and more. • Reduced dependence on foreign oil through utilizing renewable energy and clean technologies.
Reduced total costs	• Reduced hazardous management costs (e.g., using less toxic products). • Reduced operational costs (energy savings from efficient equipment). • Reduced disposal costs (hazardous and solid waste) by generating less waste and using longer lasting products. • Reduced repair and replacement costs when using more durable and repairable equipment. • Reduced costs and potential liability related to employee safety and health at the facility. • Reduced material and energy consumption. • Lower initial purchase costs.
Reputational benefits	• Enhanced organization reputation of being environmentally conscious. • Greater appeal of products/services to certain stakeholder groups.

green is expensive. However, green can save money directly as well as indirectly, as is also shown in Table 2.3. In addition, it can enhance an organization's reputation if done with good intentions and transparency.

It is important that management agrees that EP can be beneficial, or it may be looking for reasons for the EP efforts to fail. While many organizations do not want to pay a higher total cost for green, SM can emphasize EP efforts in areas where cost savings are possible.

An Approach to EP Implementation

One approach to implementing EP is presented in Table 2.4. The approach presented here works well as organizations begin to implement EP. This approach may also be effective in the long run when companies are purchasing standard items and there are several options available to choose from, and possibly multiple suppliers.

One important criterion in step 3 of this process is to consider the lifecycle impact of items that an organization buys. Concurrent with the implementation of EP, organizations should consider how they use products and then use these products efficiently with an effort toward reducing waste, as suggested in Figure 2.1. They way in which items are disposed of by the organization also has a tremendous impact on the true impact of the item over its lifecycle. Companies like Sunny D, with its Zero Waste to Landfill policy in manufacturing, have carefully considered this in everything that they buy and have developed processes and thoroughly trained their employees to reduce waste, reuse where possible, sort waste, and recycle it.

Companies that manufacture products or services, like Sunny D, can only achieve these goals because they have planned and designed their products and material choices with the end in mind. The impact of the

Table 2.4. Implementing EP[19]

1. Establish a multidisciplinary team: Key stakeholders and subject matter experts
2. Set specific objectives: Reduce landfill 10% in two years, buy 35% recycled product, etc.
3. Integrate green criteria into purchasing process: Use specific, relevant certifications, self-reported practices or audits
4. Define the purchase item and develop the specifications: Perform research to make sure what you want is available
5. Use standard labels and established criteria: Examples are the energy star certification in the United States and green mark in Taiwan
6. Choose the appropriate supplier: This may exclude suppliers who do not meet certain environmental certification, or have a record of violating environmental standards
7. Publicize and measure the success: Usually based on the objectives set in step 2 above

decisions that manufacturers make at the design phase play a major role in determining what inputs will meet their specifications, which in turn really limits the choices of input. Thus, it is absolutely essential that if these organizations want to have real impact, their environmental footprint must consider the lifecycle implications of their product design and material choices in advance.

Lifecycle Perspective

To understand and improve the environmental footprint of a product requires a lifecycle analysis approach. Understanding the environmental impact that a product or service has over its entire lifecycle is intuitively appealing, yet complex in practice. The concept of design for environment (DfE) or cradle-to-cradle manufacturing attempts to address the true environmental product lifecycle issue. An environmental lifecycle perspective addresses all aspects of a product lifecycle to get a true picture of the environmental impact of a product over its lifecycle—from design, selection, and production of components and their assembly, how the product is actually used (including energy usage and other waste generation), how long the product is used (with a long lifecycle perhaps being better), and how it is handled at the end of its life. The lifecycle perspective presents an ideal—how people use a product on average. Actual behaviors are sometimes unmanageable and can affect the lifecycle impact. For example, someone who owns a Prius but has many rapid starts and stops will not get the full benefit of the great gas mileage a Prius can have. Someone who owns a Nissan Leaf and lives in an area where the energy comes from coal will have a different environmental footprint than someone who lives in an area where energy comes from renewables. Figure 2.2 illustrates some of the major considerations at various stages of a product lifecycle.

The number of stages identified in a product lifecycle can vary depending on how finely items are divided up. Product design, which is not shown in the figure, still probably has the greatest impact. The five shown in Figure 2.2 are as follows:

1. Raw material extraction and processing
2. Manufacturing of the product

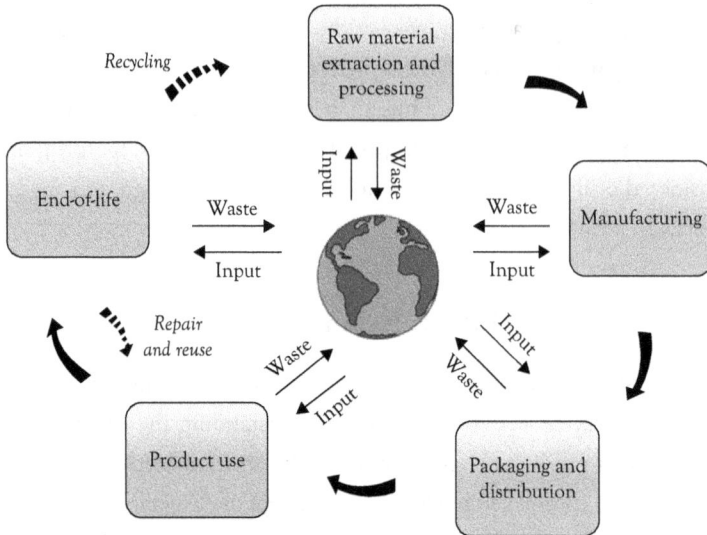

Figure 2.2. Considerations at various stages of product lifecycle.[20]

3. Packaging and distribution to the consumer
4. Product use and maintenance
5. End-of-life management: reuse, recycling, and disposal[21]

EP must be part of DfE, so it is important that supply management understand the DfE concept, and include considerations beyond whether they are buying a product made from recyclable materials to consider whether the purchased item will be able to be recycled or put to alternative use at the end of its useful life. DfE encourages a much more holistic picture of environmental impact. Hewlett-Packard has identified three priorities for its DfE efforts:

- *Energy efficiency*: reduce the energy needed to manufacture and use the products
- *Materials innovation*: reduce the amount of materials used in the products and develop materials that have less environmental impact and more value at end-of-life
- *Design for recyclability*: design equipment that is easier to upgrade and/or recycle[22]

Thus, DfE and EP both consider environmental issues beyond the immediate purchase. The cradle-to-cradle perspective fully embraces DfE, and suggests that organizations only use safe materials in production, and that those materials can be disassembled and reused or recycled at the end of the product's life.[23] There should be no waste. BMW, the leader in the Automotive Sector of the Dow Jones Sustainability Index for 2010–2012, does just that in the design of its cars. Companies who embrace DfE and the environmental product lifecycle perspective would modify the steps in the EP process to integrate DfE as part of each step of the process and focus more holistically, perhaps on customized items that could meet their performance objectives.

Supply managers also need to understand how the lifecycle approach applies to capital purchases. Capital equipment, such as production machinery and vehicles, has different initial cost (both financial and environmental), different productive capacity, different energy usage, different maintenance costs and footprint, and different productive lifecycles. The environmental impact of such equipment can be complex. In order to make sense out of such decisions, supply managers should gather all of the environmental data that they can regarding the alternatives and transform that data into comparable measures. For example, if one piece of equipment has a significantly different life expectancy and/or productive capacity, looking at these pieces of equipment side by side would not be a fair comparison. Rather, the environmental impact should be adjusted based on how the equipment will actually be used, not its theoretical maximum.

What to Measure

The lifecycle discussion above raises the issue of what sort of measurement of environmental impact is meaningful. There are key measures in several areas that many companies are focusing on. These include measures of emissions, water usage, and waste disposal. There are regulations related to the "safe" or legally allowable levels of emissions and types of emissions and waste allowed in each of these areas. However, these are really the minimum standards. Progressive organizations want to stay ahead of regulations, and reduce their environmental impact to the level possible, without creating excessive costs.

Air. Air pollution measures seem to get the most attention. One of the most accepted measures of air pollution is carbon emission, also known as CO_2, or greenhouse gas (GHG). This is not by any means a complete measure of emissions, as there are many other types of very harmful emissions, including particulate matter, The U.S. EPA has established air quality standards for five major air pollutants, as regulated by the Clean Air Act: Ground-level ozone, particle pollution (also known as particulate matter), carbon monoxide, sulfur dioxide, and nitrogen dioxide. According to the EPA, ground-level ozone and airborne particles pose the greatest threat to human health in the United States.[24] Many organizations, both for-profit and not-for-profit, exist to try to help organizations achieve carbon neutrality.[25] Tools such as carbon calculators have been developed.

Water. Water resources are important to society and to ecosystems. Society needs a clean, reliable source of drinking water to sustain health and human life. Water is also used for agricultural production, energy production, and manufacturing. The use of water puts pressure on this limited resource. These pressures are exacerbated by climate change, which is increasing demand and shrinking supplies. Water is one of these publicly provided resources and services that many people count on, and for which they have no alternative source. Coca-Cola is a significant consumer of the world's water supply. Realizing the seriousness of the water issue, the organization made a commitment to "return to communities and nature an amount of water equal to what is used in their beverages and their production."[26] To meet this goal they have three objectives: reduce the water used, recycle the water used, and replenish the water through a global network of local partnerships and projects. Other organizations are assessing their water usage and making adaptations to both process and products.

Solid Waste. The notion of zero waste to landfill is gaining attention. There are always opportunities for industries, businesses, institutions, and public facilities to reduce solid waste generation. For example, Walmart's 2012 Global Responsibility Report indicates that 80.9% of waste from its U.S. operations was diverted from landfills in 2011. In doing this, Walmart notes that it achieved cost savings through a combination of increased recycling revenue and decreased expenses.[27]

Environmental measures of supplier performance
(Electric) energy usage
(Fuel) energy usage
(Input/withdrawal) water usage
(Output/discharge) water usage
Hazardous waste disposal
Nonhazardous waste disposal
Kyoto greenhouse gas: Emissions direct (scope 1)
Kyoto greenhouse gas: Emissions indirect (scope 2)
Annual volume/output
P&G: Sustainability ideas & initiatives supported
Fines & sanctions
Environmental management system
Data protocol
Optional measures
Renewable energy
Kyoto greenhouse gas: Emissions indirect (scope 3)
Potential waste material: Recycled, reused, recovered
Transportation fuel efficiency: (Transportation suppliers only)
Industry certification
Other

Figure 2.3. Environmental measures of supplier performance for Procter & Gamble.

Cost Savings. While some companies have been criticized for viewing EP as just another cost savings approach, documenting cost savings from EP can help sell the idea within the organization. It can be an additional benefit that can raise the level of commitment and support from the C-level and from other groups within the organization. IBM has been tracking cost savings and returns associated with environmental initiatives for over a decade. It enjoyed substantial savings by reducing its electrical usage by 4.5 billion kilowatt-hours between 1990 and 2006.[28]

Procter & Gamble (P&G) is taking a very proactive approach with its suppliers with regard to environmental sustainability. They recently implemented a scorecard that measures the environmental performance of its suppliers in a number of key areas. The major categories on the P&G supplier scorecard fall into the classifications of energy, water, waste, and GHG. An example of the metrics tracked on the scorecard is shown in Figure 2.3.

Like P&G, IBM believes that real results in GHG emission reduction are directly achieved when each enterprise takes responsibility to address its own emissions and improve its energy efficiency. This is reinforced by IBM's recent announcement that all of its first-tier suppliers will be expected to develop a management system, inventory their key environmental impacts including GHG emissions, and develop reduction plans for those key impacts.[29] Hence, they introduce how important it is for the purchasing function to work with its supply chain and supply base.

Key Points

- There is a continuum of effort related to environmental purchasing, from buying items made from renewable resources through considering the environmental impact of an item over its lifecycle. Some organizations start with buying renewable items and progress from there. Others remain at that basic level due to limited resources or limited impact associated with greater effort.
- The waste pyramid was introduced to reinforce that materials no longer fit for their initial use can be managed in very different ways, each with different environmental impacts.
- There are many benefits that an organization could achieve through EP implementation, from reducing its footprint to lowering waste and associated cost.
- A basic seven-step approach to implementing EP was presented. This approach suggests that EP should be a team effort, and the depth of EP implementation must fit the organization's resources and goals.

- The product lifecycle concept was presented to reinforce the importance of taking a holistic approach when analyzing decisions related to the environment. Different products have very different impacts at various stages in their lifecycles.

- When implementing or considering EP, it is important to measure EP performance and outcomes. These measures can be integrated into supplier selection and ongoing evaluation. Such environmental metrics may include areas such as land, water, air, and overall cost improvement. Actual supplier performance should be measured and tracked.

Internally, there are various breadths in the scope of pollution that organizations track, from internal only (scope 1) to the whole supply chain (scope 3).

CHAPTER 3

What Are Some Key Concepts in Understanding Green Sourcing and Its Impacts?

Introduction

There is a continuum of depth in the way that organizations can view sourcing for the green initiative. This includes a very basic perspective: from buying products that are made from recycled materials, to conceiving and designing them holistically, and integrating all aspects of environmental impacts over them in their lifecycles. The impact of the items the supplier provides to an organization has implications for the way in which a supplier relationship should be managed, at what stage of the product/service development process a supplier should become involved, and the ongoing relationship with the supplier. In light of this, this chapter opens with an introduction to the concept of supplier segmentation for the environment, and how and where environmental issues fit in to the supplier selection and management process. The nature and the role of supplier certification and audits in EP are then examined. The concept of reverse logistics and closed-loop supply chains is presented for consideration as another way to improve EP and sustainability in certain applications.

Segmenting the Supply Base for the Environment

Supplier segmentation is a critical concept for the purchasing function. It was popularized by Kraljic's purchasing portfolio matrix. Kraljic suggests that purchases be segmented based on the importance of the purchase and the market complexity associated with a purchase. Considering the importance of the purchase, this framework considers the cost of materials, the profit impact, and other significant ways the purchase might affect the organization's success. Complexity of the supply market includes issues such as competitiveness, availability of viable sources, pace of change of technology, logistics, and other factors that make the supply market a challenge to deal with.[1]

Combined, these characteristics in turn determine the way a purchase should be managed, including the ideal type of relationship that the organization should have with the supplier of an item. A brief version of Kraljic's purchasing portfolio matrix is presented in Figure 3.1. This illustrates that there are four major types of purchases: leverage, strategic, bottleneck, and noncritical, and the supplier of each commodity should be appropriately managed. It is important to note that these classifications are not static. As market conditions and firm's priorities change, the way in which purchases are classified needs to be continuously revisited. For example, if an item that is categorized in the leverage category suddenly has new technology introduced that only a few suppliers are capable of providing, then that item would likely shift to a more strategic classification. Or if a raw material that was relatively easy to obtain such as lumber

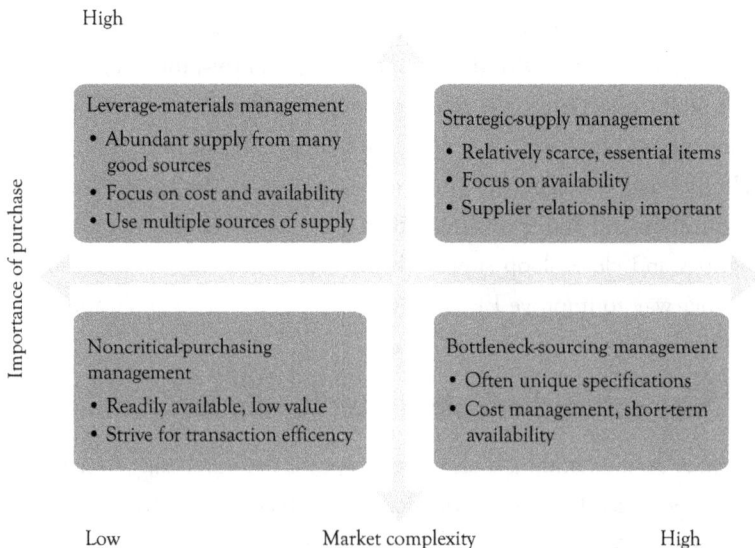

High

Importance of purchase

Leverage-materials management
• Abundant supply from many good sources
• Focus on cost and availability
• Use multiple sources of supply

Strategic-supply management
• Relatively scarce, essential items
• Focus on availability
• Supplier relationship important

Noncritical-purchasing management
• Readily available, low value
• Strive for transaction efficiency

Bottleneck-sourcing management
• Often unique specifications
• Cost management, short-term availability

Low   Market complexity   High

Figure 3.1. Purchasing portfolio matrix.[2]

suddenly has supply availability issues due to regulation or natural scarcity, this would shift to a more strategic or potentially bottleneck category until the supply market restabilized.

Expanding on Kraljic's concept of segmentation, not all purchased items should be treated equally based on their potential impact on environmental sustainability. Pagell et al. (2010)[3] suggest that purchases should be segmented by simultaneously considering the supply risk associated with the purchase, and the threat that the purchase poses to the organization's triple bottom line, as illustrated in Figure 3.1. They propose that items that are a high threat to the organization's sustainability and a high supply risk (strategic item) be given the most attention. The sustainability dimension includes environmental, social, and financial implications of the supplier's practices. The strategic category includes the purchases where the supplier's conduct needs to be most fully understood and monitored, because the risks are the greatest.

Similarly, Krause et al. (2009)[4] suggest that sustainability be added as a key performance criterion in Kraljic's matrix. They offer some suggestions regarding how additional criteria related to sustainability might vary depending on the type of purchase. Expanding on and modifying this suggestion, environmental issues can be added as a criterion that affects the dimension "Importance of the Purchase." This could be implemented in a number of different ways. The purchased item would be classified

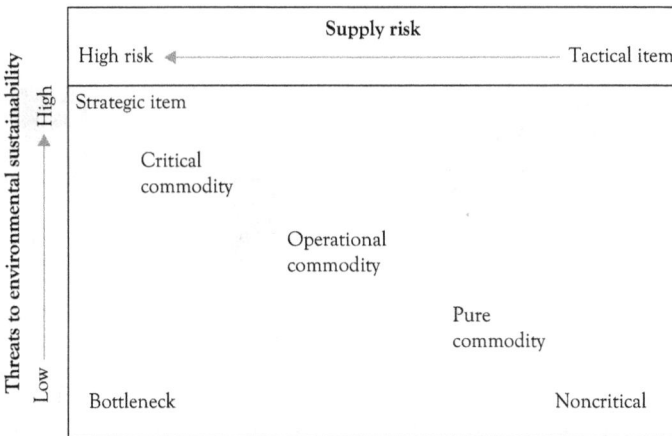

Figure 3.2. The sustainable purchasing portfolio matrix.[5]

based on the usual criteria from the purchasing portfolio matrix in Figure 3.1. Key methods for addressing EP could be incorporated into each quadrant. This is illustrated in an adapted matrix in Figure 3.3.

It is possible that in some cases the nature of potential contribution of an item to the organization's total footprint could be so great that the purchased item should actually be moved to a different quadrant. For example, strategic suppliers are those who provide items that are of high supply complexity, whereas leverage items have many viable sources. If an organization decides to really focus on cutting emissions and works with a "leverage" supplier to change its formulation to one that is much less polluting, that supplier may begin to move into the "strategic" realm, and be included in new product development, and no longer be viewed as easily substitutable. Thus, purchased items that have greater environmental impact may be elevated to a different type of supplier management as the organization works to reduce the impact of those items.

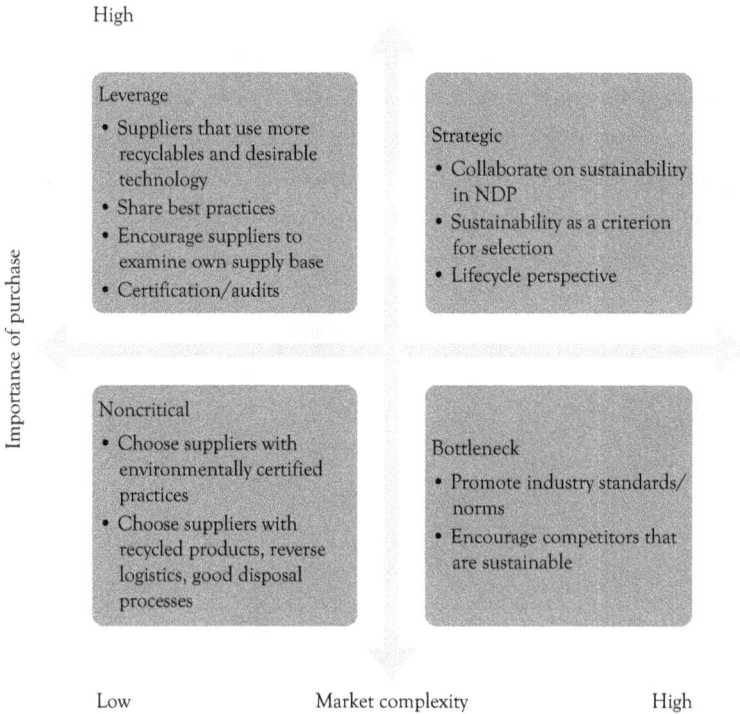

Figure 3.3. *Key supplier selection and management issues based on EP.*[6]

Items can also shift from the bottleneck category to one that is more important or strategic to the organization. A raw material fabricator had a number of environmental projects: soil testing, measuring of GHG, and more. A different consulting firm was hired to support the firm for each project. One consulting firm started to specialize in environmental projects, and thus received additional projects (and therefore revenue), as well as a deeper understanding of the firm's processes, which shifted that supplier to a more strategic relationship, one that required a different type of management and oversight.

EP in the Strategic Sourcing Process

An essential element of the strategic sourcing process is supplier segmentation. But the strategic sourcing process has broader implications for supplier selection and management that should be adapted to embrace EP. A figure of a typical strategic sourcing process is shown in Table 3.1.

While this provides an overall framework, the approach used will vary based on the way that those involved in the strategic sourcing process answer the questions. To illustrate this, IBM has long embraced the notion that suppliers should be treated differently based on their potential impact on the environment. IBM had documented, approved policies related to "IBM Environmental Requirements for Materials, Parts, and Products," in the early 1990s, and even before that time. These specifications had objectives "… regarding IBM products and materials used in IBM products:

- Conserve natural resources by reusing and recycling materials, purchasing recycled materials, and using recyclable packaging and other materials.
- Develop, manufacture, and market products that are safe for their intended use, efficient in their use of energy, protective of the environment, and that can be reused, recycled, or disposed of safely."[7]

At that time the focus was on clarifying which substances were prohibited from use, clarifying documentation of substances that required disclosure, and proper handling and disposition of any potentially hazardous materials, such as batteries. [8]

Table 3.1. EP Criteria Incorporated into Strategic Sourcing

Process Step	1.0	2.0	3.0	4.0	5.0	6.0	7.0
	Determine needs or opportunities	Assess possible approaches	Develop strategy	Screen and select suppliers	Solicit RFPs or bids	Negotiate and establish agreement	Manage and measure continuous improvement
EP considerations	How critical is this item to our environmental footprint? Is this a hazardous material?	What EP factors should we incorporate into our minimum standards for screening suppliers? What are key EP selection criteria?	In which quadrant does this item fit, and what approach is appropriate based on that? How many viable choices do we have?	Type of EP prescreen based on self-audit, external environmental certifications, and possibly audit finalists? Suppliers meetings needed before solicitation?	Fit with EP criterion? Does this supplier offer better ideas reducing our environmental footprint? Site visits of finalists needed?	How to include EP requirement in agreement? Any incentives or cost sharing support for environmental improvements?	How to measure and report EP performance, continuous improvement expectations, and achievements?

Looking at the strategic sourcing process, IBM segments suppliers in the very first step of the process based on the nature of what they supply. It has a clear set of packaging guidelines that are applied to packaging suppliers. In addition, suppliers that provide chemicals or suppliers that assist IBM in disposal have been subject to very stringent selection criteria as well as audits for more than 20 years. Suppliers who cannot meet IBM's criteria are disqualified from consideration. EP is incorporated into the remaining steps of the strategic sourcing process based on how the incorporation of environmental requirements was established in the initial step.

IBM's approach to EP continues to grow to one of sustainable supply chain management. In 2010, IBM implemented Social & Environmental Management System (S&EMS) globally. This encompasses the following elements:

- Define, deploy, and sustain a management system that addresses corporate responsibility, including social and environmental stewardship
- Measure performance and establish voluntary, quantifiable environmental goals
- Publicly disclose results associated with these voluntary environmental goals and other environmental aspects of their management systems
- First-tier suppliers to cascade these requirements to their own suppliers[9]

IBM is implementing an assessment and re-audit process to ensure compliance. The topic of supplier assessment and audits is becoming key for those involved in managing the supply base. More information about supplier assessments and audits is presented in the next section.

Supplier Assessments and Audits

The purpose of a supplier environmental audit is to determine their behaviors and activities, as well as compliance to environmental regulations. There are external organizations that conduct these to meet specific certification criterion, such as ISO 14001. Some industries have common audits and some companies individually audit their suppliers. There are pros and cons to each approach. The ISO 14001 certification, developed by the International Standards Organization, is based on an audit of a company's processes, and certifies that the company retains documentation on procedures and processes that have an impact on the environment. Much like the ISO 9000, this is an extremely well-recognized and voluntary certification. The ISO 14001 certification audits processes related to air, water, solid waste, energy, and other environmental impacts.[10] It does not ensure environmental responsibility, only that an organization documents what it does. The ISO 14001 has been highly criticized for its inability to ensure compliance. As a result of this criticism, many more environmental certifications and audits have developed. Numerous for-profit and not-for-profit organizations that specialize in supplier environmental certification and audit have started to gain traction and influence with purchasing and others that work with supply base management.

Much like a supplier quality audit, a supplier environmental audit process generally involves some sort of self-assessment, which includes completing a questionnaire about environmental behaviors, emissions, systems, and tracking and returning this to the audit team before the location visit. These audits frequently include other aspects of sustainability, such as labor practices. This self-assessment is generally reviewed by the audit team before the audit team visits the supplier's facilities. An audit team often includes those with technical and business expertise. The team may be the customer's employees, or a qualified third party. For example, Apple's supplier audit teams are led by an Apple employee, and supported by local third-party auditors who have expertise in a given area.[11] If the audit is performed by a third party, the customer may pay the expense, or it may require that the supplier pay. The latter has been common with ISO certifications, as they are seen as very general certifications that are the supplier's responsibility. Frequently, ISO certifications are considered a common order qualifier, particularly in the EU.

Once on-site, the audit team may spend several days walking through processes, procedures, and documentation. It generally examines the management structure for proper segregation of duties and responsibilities, as well as reviews training programs. This on-site audit allows a deep and broad understanding of the supplier's capabilities. The supplier will receive the audit results and may be required to change its practices immediately or to phase in improvements. If things are out of compliance, it is possible that a supplier could be suspended by a customer. For example, Apple conducted 229 supplier audits in 2011. It discovered 112 facilities that were not properly handling hazardous chemicals. It required them to establish procedures for proper chemical handling and storage, and helped them where necessary.[12] Corrective action and clear timelines for improvement are essential outcomes of unacceptable audit results.

One organization that provides a uniform audit process is the Electronics Industry Citizen Coalition (EICC). This voluntary electronics industry group has developed a code of conduct encompassing member behavior related to labor, health and safety, environment, ethics, and management systems.[13] It has developed self-assessment tools and audit tools to report on conformance. There are also third parties that are certified as auditors. Organizations that are EICC members generally are willing to accept an EICC audit result provided by another organization rather than audit the supplier itself. This can save time and money of both the buyer and the supplier firm. These audits, while very worthwhile, are very labor-intensive and costly. Where it makes sense, customers should share audit results related to a specific supplier or provide some type of certification that negates the need of performing redundant audits on a supplier.

Environmental audits have been criticized for providing only a snapshot in time and not focusing on continuous environmental improvement. If the audit is viewed as an event, rather than part of a larger process, that may be true. However, there are an increasing number of companies, such as Nokia, that focus on commitment from their entire supply chain in ensuring environmental improvement. Nokia takes a first-tier approach in supply chain management, aiming to positively influence the first tier of suppliers and to encourage them to spread responsibility throughout the entire supply chain. Although the approach is time-consuming, it is effective because it is sustainable in the long term.[14] In the electronics industry specifically, supplier relationships are often long term and the supply base can be quite stable. Nokia works closely with its suppliers, which provides a good opportunity to improve their environmental performance as needed. These suppliers have additional pressure to improve environmental performance, including increasing legal requirements, and other important customers who are also setting stringent environmental performance requirements.

Reverse Logistics and Closed-Loop Supply Chains

Performance metrics and the environmental supplier audit often include a process for disposal or reuse of products once they have completed their useful life. This is often discussed as a "take back" program. Both manufacturers and retailers are being held accountable for developing programs where products are returned at the end of the useful life and then recycled, remanufactured, reused, or properly disposed of. This is still considered "voluntary" in most states within the United States. However, this process has become highly regulated in the EU. Regardless of regulations, more and more companies are taking responsibility for a product's disposition at the end of its life.

Also, as discussed in Chapter 2, companies need to include environmental criteria within the product design stage. Organizations should consider the entire product lifecycle, including disposition when the customer has finished using the product. In an interview with *Atlantic Monthly*, Kevin Dooley, Director of The Sustainability Consortium, reminded us, "Most people are largely unaware of the life cycle of the products they purchase, and their environmental and social impacts. They worry about the packaging and whether it's recyclable, which is good, but are unaware of the effects related to the product's supply chain, from raw material extraction to manufacturing to distribution and retailing."[15] How product inputs and suppliers are selected to manufacture the products has a huge impact on the final product and also affects disposition at the end of the product life. This becomes an essential consideration in strategic sourcing for EP. If the product is designed so that it is easy to disassemble, and the materials are able to be reused, remanufactured, or recycled, then the entire product end of life becomes easier to manage and less environmentally damaging.

There is a directive in the European Union, known as the "Waste from Electrical and Electronic Equipment" (WEEE). This directive has changed the way that manufacturing and sourcing are performed. Beginning in 2000, the WEEE directive required manufacturers of cars and electronic products to provide disassembly centers. This included the development of a "reverse logistics" process to get the item back from customers when the customer is done using it. BMW had already begun to design its cars for the environment (DfE), meaning that its cars were made for ease of disassembly, and included materials that were easy to reuse or recycle, rather than have to be put in a landfill.[16] This requires careful specification of plastics that can be recycled and have value in the after market. These considerations are essential in the supplier selection process.

Thus, if the items that you are buying affects the delivery and return process for the product that you produce, this needs to be understood in the first step of the strategic sourcing process, and factored in to the selection criterion in step 4 (see Table 3.1). Transportation emissions are a significant aspect of most supply chains' environmental footprint. The concept of "reverse logistics management focuses on the movement of a returned product to recapture value by quickly making decisions on how the product will be managed and re-enter the supply stream."[17] There are two general types of classifications of this return activity. One is reverse logistics, which loosely describes all of the processes and execution of activities related to return and disposition of materials, products, components, and packaging.[18] The other type is a closed-loop supply chain, in which the product has been designed specifically for disassembly and reuse.

There is often a "natural" way that the customer returns the product at the end of its life that is associated with replenishing the product. Examples of this include empty beer kegs, which bars return when they purchase full kegs, and commercial tires, which are often retreaded and resold after they are removed and replaced at a truck service center. There are hybrid types, such as print cartridges, which may be returned to an office supply store when you purchase a new one, and will either be refilled or recycled. The key is that there is a formal process in place that is very easy, often seamless, for the customer to deal with. Also, because the product is designed with returns in mind, as mentioned above in the BMW example, the recovery from these returns is usually higher, with less waste and landfill materials generated.

From an EP standpoint, reverse logistics can also consider the delivery and return process for the items that you are buying. For example, companies like Honda of America Manufacturing have parts packaged in reusable totes. The empty totes are returned to the supplier the next time that a pick is made. This saves a large amount of cost and packaging waste. The return is just part of the regular pick-up process. This is a closed-loop logistics system. This type of requirement can be part of the supplier selection process. It is also much more efficient if the supplier is located nearby and has regular shipments, so that reusable container inventory does not build up. Thus, supplier location would also become a more important selection criterion in this type of system. This just touches on the surface of the reverse logistics process as they relates to EP. For more information, check the Reverse Logistics Executive Council website at www.rlec.org.

Summary

- Environmental issues can be incorporated into the strategic decision-making of the purchasing department at each step in the strategic sourcing process.
- Environmental projects and initiatives can influence the nature of the buyer–supplier relationship. In some cases, the environmental activity can cause a shift in the importance of the relationship and therefore the amount and nature of attention that the relationship receives from the buying organization.
- Supplier environmental assessments and audits can be an effective way to monitor supplier performance and identify areas for supplier environmental improvement.
- Reverse logistics and closed-loop supply chains consider how products are handled at the end of their useful lives and require supplier cooperation.
- As supplier relationships change and become more integrated and strategic in fulfilling the organization's objectives, a longer-term perspective must be initiated to meet the needs of and value to both buyer and supplier organizations.

CHAPTER 4

What Industries and Sectors Lead the Way: Part 1

Introduction

The first chapters in this book have provided an overview of some of the key practices in EP. This chapter provides some insights into industries and companies that lead the way in environmental purchasing practices and working with their suppliers. Examples of how good intentions can also have negative impacts are provided.

This chapter focuses on some of the strengths and weaknesses in the environmental practices of the electronics and automotive sectors, relating to EP practices. It begins with a discussion of some of the environmental regulations implemented by the European Union (EU) and how leading companies have responded. The first regulation addressed the challenging issue of electronic waste (e-waste), how the electronics industry has attempted to deal with it, the problems it has faced, and the actions it has taken to address those problems. Next, it will move to the issue of end-of-life disposal, emphasizing the concept of environmental issues over the product lifecycle.

Regulations and the Rationale Behind Them

As mentioned earlier, most companies still only consider Scope 1 waste and emissions when reporting, and when designing their products and processes. But because so many companies and activities outside of the original equipment manufacturer's (OEM'S) factory walls contribute to GHG and other forms of waste, to consider only Scope 1 emissions paints a very limited picture of the real problem. In many cases, the manufacturer designs the product, the specifications, and material inputs. The manufacturer of the product also chooses the transport mode and the packaging, and thus both directly and indirectly contributes to Scope 3 emissions—the entire emissions of the supply chain for a product, including the product's use. Much of the legislation implemented in the EU is in direct recognition of the fact that manufacturers need to consider and take broader responsibility for the impact of their products across the lifecycle.

Restriction of Hazardous Substance (RoHS) and Related Regulations

The regulation of hazardous materials is very complex and governed by many local, national, and even global laws. The notion that various materials and chemical compounds could hurt, permanently damage, and even kill people and animals really started to get attention with the publication of the book *Silent Spring*[1] in 1962, which exposed how various pesticides, particularly dichlorodiphenyltrichloroethane (DDT), were destroying wildlife's ability to reproduce. This made the general public acutely aware of the potential dangers of not understanding what and how various materials are used. In an attempt to unify much of the knowledge and regulations related to hazardous materials, the EU introduced the RoHS in the early 2000s.

> The RoHS directive aims to restrict certain dangerous substances commonly used in electronic and electronic equipment. Any RoHS compliant component is tested for the presence of Lead (Pb), Cadmium (Cd), Mercury (Hg), Hexavalent chromium (Hex-Cr), Polybrominated biphenyls (PBB), and Polybrominated diphenyl ethers (PBDE).[2]

This regulation applies to anyone who wants to manufacture, sell, or import to the EU. It covers the following categories:

- Large household appliances
- Small household appliances
- IT and telecommunications equipment
- Consumer equipment
- Lighting equipment

- Electrical and electronic tools (with the exception of large-scale stationary industrial tools)
- Toys, leisure, and sports equipment
- Automatic dispensers[3]

Within the United States, similar legislation has been passed at the state level in California and Minnesota, and is under consideration in many other states. There are additional regulations dealing with other types of hazardous materials that vary by state. In addition, there are a host of regulations related to transporting hazardous and volatile materials. It is the responsibility of supply management to ensure that it is compliant with all of the relevant regulations for all of the materials that it purchases. It is also the responsibility of supply management to ensure that its suppliers are compliant with the relevant regulations. As the laws continue to change, it is helpful for supply management to be aware of materials that are currently under scrutiny and work to have those items eliminated from its products and its purchases from suppliers.

Waste Electrical and Electronic Equipment Directive

Another significant EU directive is the Waste Electrical and Electronic Equipment Directive (WEEE). Like RoHS, it applies to companies that want to manufacture, sell, or import to the EU. The goal is to sharply reduce the amount of electronic waste (e-waste), especially hazardous materials that end up in landfills. It is complementary with RoHS. It requires that manufacturers inform consumers of their products' end-of-life options, cover all the costs of safe product recovery and disposal, and report on their activities.[4] It was approved in 2003 and implemented in 2005 but continues to evolve. Interestingly, a U.S. computer company, Hewlett-Packard (HP), is a founding member of the European Recycling Platform, which established the framework for WEEE compliance. HP states that it has been "designing for recycling disassembly" since the early 1990s. This includes measures such as the following:

- Use of common fasteners throughout a product
- No more adhesives

- Use of more metals, which are more easily recyclable than plastics
- Use of plastics families, rather than chemically disparate plastics, such as aromatic-based and olefin-based
- More snap fits
- Clear markings indicating types of plastic used[5]

In recognition of the need to be proactive in addressing environmental issues, a group of primarily computer companies developed the voluntary Electronics Industry Citizenship Coalition (EICC),[6] which developed the Electronics Industry Code of Conduct for working with its suppliers. The code of conduct was drafted in 2004 by HP, IBM, Dell, and several other firms in order to create standards for working conditions and environmentally responsible manufacturing practices. In addition to addressing the issues laid out in the RoHS and WEEE directives, a major goal of the EICC was to create industry-wide harmonization of approaches to dealing with supplier performance in regard to social responsibility, including human resources, ethics, and environmental issues. A key benefit of this approach is "[b]y consolidating and standardizing compliance, audit and reporting efforts, suppliers can focus on achieving the high standards of performance set forth by the Code".[7] Thus, this effort recognizes the benefit of "shared efficiencies" throughout the supply chain, rather than just compliance. EICC membership has broadened, and is shown in Figure 4.1. Many of the companies listed are suppliers to the electronics industry. It is extremely beneficial to have them all following the same set of standards. Not only is it much more efficient to follow one set of standards, a common set of standards clarifies expectations of the suppliers, reduces costs of auditing and compliance to multiple standards, and really creates a common culture around environmental practices and the broader issue of sustainability.

According to a recent study, the companies that comprise the environmental leaders in the electronics sector are the most proactive sector in terms of involving their suppliers in environmental initiatives and really engaging members of the supply chain in reducing their environmental footprint.[8] Several examples from the members' environmental reports help describe the specific areas of engagement for the leading EP adopters in the technology sector.

Figure 4.1. EICC members.[9]

In 2006, Dell noted, "Dell recognizes the responsibility (it has) to work with (its) suppliers to promote sustainable social and environmental practices. (It) also understands that (it) can deliver more efficient and consistent results through cooperation and coordination among the Electronics Industry Code of Conduct (EICC) partners."[10] In 2012, it reported, "Dell's supplier capability-building activities uphold suppliers' internal ownership for social and environmentally responsible behavior; they are designed to enable Dell partners to make changes for long-lasting and sustainable impact. Dell hosted more than 100 suppliers at capability-building workshops in FY12. Dell (also) conducted 125 facility audits of its suppliers (including 25 Electronic Industry Citizenship Coalition (EICC) third-party validated audits) and released consolidated audit findings about areas of non-conformance necessitating corrective action."[11] Dell's report suggests that it has moved beyond compliance to supplier engagement. While Dell has established formal supplier principles related to sustainability, suppliers are also well integrated at each stage of Dell's lifecycle management philosophy of environmental responsibility. Dell has established Design for Environment, whereby suppliers are engaged in everything from making designs more energy efficient to making sure materials meet or exceed environmental standards. Dell works with suppliers to identify opportunities to substitute more environmentally friendly materials, and audits suppliers' specifications, materials, and processes.

Hewlett-Packard (HP) takes a risk-based approach to Supply Chain Social and Environmental Responsibility (SER), regularly auditing and focusing its attention on those suppliers that it believes pose the greatest potential sustainability risk. In terms of supplier development, HP trains its first-tier suppliers to audit its second-tier suppliers, in order to permeate its SER philosophy throughout the supply chain. In addition, HP works with its first-tier suppliers to train the second-tier suppliers to better manage the environmental initiatives of its suppliers. An example of this is HP's "partnership" with one of its key suppliers, Flextronics. HP and Flextronics work together to organize SER supplier forums, helping suppliers learn from each other. Flextronics notes that HP has taken a unique approach by "... influencing suppliers to join its cause."[12] HP has conducted over 700 supplier EICC audits and uses the following key metrics related to supplier sustainability:

- Suppliers engaged in SER program
- SER audits conducted
- Distribution of audit findings
- Nonconformances reduced by section of HP's EICC Code
- Distribution of major nonconformances by section of HP's EICC Code
- Suppliers engaged in capability building[13]

As such, HP has taken a leadership role in the Electronics Industry Code of Conduct (EICC) effort. HP notes that, "We support individual producer responsibility (IPR), and believe that all manufacturers should share responsibility for managing electronic waste with governments, retailers, and customers."[14]

HP also recognizes the reputational impact of its supply chain practices. HP's CEO noted, "... global citizenship is the 'hidden component' in HP products—embedded in our design and engineering, including accessibility, energy efficiency and recycling. It's also an important part of how we operate our business, from responsible supply chain management to the steps we're taking to reduce our own environmental footprint."[15]

In summary, the technology sector is at a significantly more advanced level of EP, whereas firms in the financial sector have done little in this area.[16] In the financial sector, environmentally oriented regulatory pressure is minimal; energy usage, perhaps the biggest impact of financial services suppliers, is invisible to the customer so there are limited reputational risks; and financial service firms have not traditionally focused on their physical infrastructure as being a source of costs and thus efficiencies.[17] The automotive sector and end-of-life directive are discussed next.

End-of-Life Directive

Like the WEEE directive, the end-of-life vehicle recycling directive in the EU recognizes that disposal of vehicles at the end of their life creates hazardous waste and adds volume to increasingly scarce landfills. This end-of-life disposal focus really began the start of a new era in waste management legislation related to durable goods on a global scale. Japan adopted similar legislation just after the EU.

While about 75% of a vehicle was recyclable at the time this directive was initiated in the early 2000s, the goal is to have 95% of the vehicle recyclable by 2015.[18] For example, the two million vehicles that reach the end of their life each year in the UK are classified as hazardous materials until they are disassembled. These vehicles are to be disassembled by authorized treatment facilities, where the public can return vehicles without charge. This creates a reverse logistics process that is becoming institutionalized in the EU. The retired vehicles are treated to remove harmful components such as fuel, oil, and batteries, while the rest of the vehicle is recycled, reused, or otherwise properly disposed of.[19]

The rationale behind this and similar legislation is Extended Producer Responsibility (EPR) for supply chain waste.[20] This is very similar to HP's notion of individual producer responsibility, where each manufacturer takes responsibility for a product it manufactures and all of its components from idea inception to end of useful life. Ultimately, the greatest improvement and reduction in supply chain waste can be made through design for disassembly, a practice that BMW began to implement in the early 1990s with the 1991 BMW Roadster. Whether affected directly by the legislation or not, there has been some movement toward design for disassembly

Major initiatives in product recycling (3 R's) over the material cycle				
Information from production, use, and end of product life feeds into the development process		New product development	Manufacturing process	Consumer use	End of life
Reduce	3R pre-assessment system	Design for reduction		Improved gas mileage	
Reuse		Design for reusability, recyclability	By-product recycling (see more)	Recycled/reused parts	Integrated motor assist battery recycling
			Bumper recycling		
Recycle		Reduction of environmental impact substances			Compliance with the end-of-life vehicle recycling law
					Voluntary motorcycle recycling

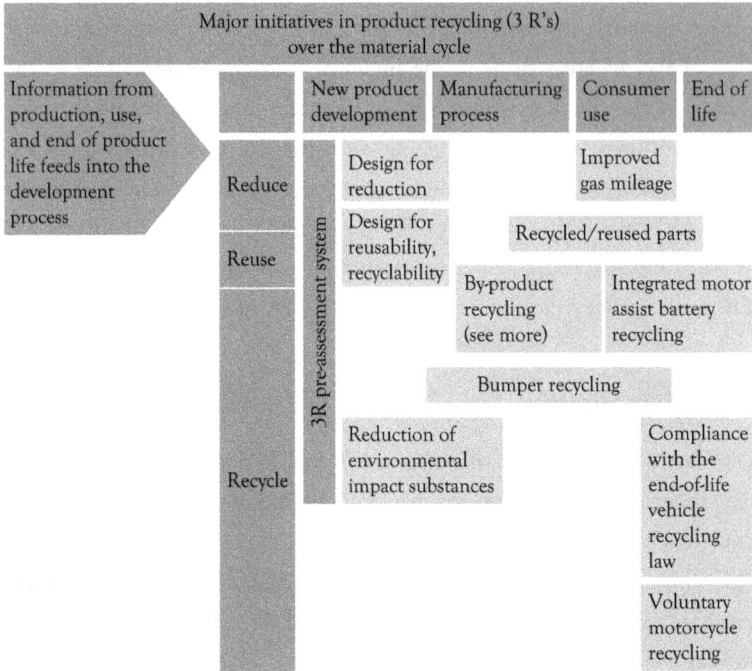

Figure 4.2. Honda's lifecycle perspective on product recycling.

initiatives by leading automotive companies such as Honda[21] and Lexus.[22] Figure 4.2 shows Honda's perspective and major initiatives related to product recycling. Design plays a very crucial role during the development process.

Long before end of vehicle life EVL legislation in 1991, Chrysler, Ford and General Motors formed the collaborative Vehicle Recycling Partnership (VRP), which is now a part of the United States Council for Automotive Research (USCAR).[23] This initiative has worked to improve techniques and materials to help design vehicles for disassembly, using easily recyclable materials, best recycling and separation practices, and even best practices in extraction and manufacturing of inputs to the automotive sector. Today, it is working on improving the disposal of hybrid vehicle. The VRP shares its findings with the industry. Today, more than 95% of vehicles are recycled in the United States.[24]

The selection of suppliers and materials are essential to the entire disassembly and disposal process. Items that may seem small to those unfamiliar with DfD, such as the type of fasteners used and the materials from which the fasteners are made, are essential to the success of DfD. Thus, it is critical that supply management understand specifications and make the requirements and the rationale for the requirements clear to suppliers and potential suppliers. Even beyond that, in its boundary-spanning role of working both with internal functions and external suppliers, supply management can help identify its most innovative and cooperative suppliers to work with engineers and designers to share their ideas to improve the product design and disassembly so the product had the minimal end-of-life footprint. These early design decisions have a huge impact on the Scope 3 emissions for a particular product, and suppliers, who are a major source of Scope 3 emissions, can suggest ways to improve the design, materials, and processes to reduce this footprint. In fact, as the Honda example of Scope 3 emissions from Chapter 3 illustrated, the biggest impact of cars on the environment is from driving the cars. Thus, designing cars with the best possible gas mileage and emissions profile should be the focus of the automotive sector.[25] However, there are some externalities in these return initiatives that were not originally anticipated.

E-Waste

Electronic waste, also known as e-waste, represents all of the electronics that are purchased that end up in landfill. It includes items such as cell phones, laptops, and televisions, many which have a life of 18 months to three years, and even less, before they are replaced. The level of waste has been growing about 8% per year and represents the hazardous materials banned by the RoHS directive presented above.[26] Some companies, like Hewlett-Packard and Dell, include shipping labels when you purchase a new computer that allow you to return your old computer to them for free and safe disassembly and disposal. Apple runs its own disposal operations. When managed this way, the waste handlers can be audited, properly certified, and monitored.

In many situations, the original equipment manufacturer (OEM) is not connected with the recovery of the goods at the end of life. In addition, computers are dropped off when a community hosts a day to drop off old electronics and appliances. Unfortunately, not all e-waste is properly handled. The problem with this is the fact that, "The toxic materials in

Table 4.1. *Management of Used and End-of-Life Electronics in 2009*

	Ready for End-of-Life management (million of units)	Disposed (million of units)	Collected for recycling (million of units)	Rate of collection for recycling (by weight)
Computers	47.4	29.4	18	38%
Televisions	27.2	22.7	4.6	17%
Mobile Devices	141	129	11.7	8%

electronics can cause cancer, reproductive disorders, endocrine disruption, and many other health problems if this waste stream is not properly managed. Many of the toxic constituents are elements, which means they never disappear, even though they may change form. Other toxic chemicals in electronics do not break down over time and instead, accumulate in the food chain and biosphere. Not only do these toxins present risks to communities and the global ecosystem, but also to electronics recycling workers, even in developed countries."[27]

Unfortunately, unlike the automotive sector, it is estimated that only about 11–14% of e-waste in the United States is recycled.[28] The latest figures from the EPA are shown above.[29]

Regrettably, cell phones, which have the shortest life expectancy, are rarely recycled. In addition to the hazardous waste generated, there are some useful materials that can be extracted from these devices in very small quantities. This makes it attractive to disassemble and recover the valuable materials if the cost of recovery is very low. In the developed world, where wages are relatively high and working conditions carefully monitored, it is generally not cost-effective to remove these materials.[30] The problem with this is these products are frequently shipped to emerging markets, such as China, India, Pakistan, and Africa, where the wages are low, and employee safety is often not regulated. It is estimated that as much as 70% of e-waste that goes to private recyclers actually ends up in another country for disassembly.[31] With improper disassembly, irreparable human health risk and environmental damage can occur. There are numerous loopholes in the laws, and tracking and enforcement are difficult and costly. The ideal situation is to reduce the amount of RoHS used in producing these items. A second alternative is to make disassembly very easy and cost-effective, but the initial product cost would likely be higher. While there are many other alternatives, very few people and companies currently think about how they are going to dispose of electronics when they purchase them. That is the topic of the next section. The black market for e-waste and its toxic and damaging effects is an unfortunate side effect of consolidating e-waste for proper disposal. Fortunately, there have been successful prosecutions of offenders, with the ultimate goal of stopping improper export and disassembly altogether.[32]

Electronic Product Environmental Assessment Tool

The EPA recommends using smarter procurement practices when buying electronics. To support this, it developed the Electronic Product Environmental Assessment Tool (EPEAT) program in conjunction with the Green Electronics Council. EPEAT provides an easy way for people to understand the relative environmental attributes of various electronics.[33] It can be very helpful to supply managers who purchase electronics. It provides a product registry that publishes ratings based on the categories of criteria listed in Table 4.2.

Items receiving "Bronze rating" meet all mandatory criteria, "Silver" meets all required criteria and 50% of the optional criteria, and "Gold" meets all required criteria and 75% of the optional criteria. While the general categories of criteria are shown in the table, there are very specific

Table 4.2. Key Categories of Criteria Evaluated by EPEAT

• Reduction/elimination of environmentally sensitive materials
• Material selection
• Design for end of life
• Product longevity/life extension
• Energy conservation
• End-of-life management
• Corporate performance
• Packaging[34]

criteria and protocols for different products to achieve different ratings, as detailed on the EPEAT website.[35] As the website lists products available in more than 40 countries, the reach of the manufacturers can extend beyond the United States. It is part of the EPA's environmentally preferable purchasing portfolio.

Key Points

- The EU has led the way in implementing legislation to require companies to take responsibility and reduce hazardous materials and landfill that result from the production, use, and disposal of their products.
- The electronics and automotive sectors have been the most affected by EU legislation.
- The electronics and automotive sectors lead the way among industries in implementing best practices in EP and supply chain management, in an effort to reduce and improve management of e-waste.
- Key EU legislation includes Restriction of Hazardous Substance (RoHS), Waste Electrical and Electronic Equipment Directive (WEEE), and the End Of Life Directive, including Extended Producer Responsibility
- The improper disposal of hazardous materials, many of which include electronics and automobiles, has devastating effects on the environment and human health.
- Design for environment is a critical concept used in the electronics and automotive sectors to understand and reduce the footprint of the entire supply chain.
- Suppliers can be engaged in the various ways of improving a company's environmental footprint, some voluntary and some regulated.

CHAPTER 5

What Industries and Sectors Lead the Way: Part 2

Introduction

Industries like automotive and electronics industries have attempted to address environmental issues by looking at the end-of-life disposal aspects of their products. This has involved rethinking the entire product design process, material selection, and even disassembly. Supply management and suppliers play a critical role in making such efforts a success. This section explores what consumer product firms and retailers are doing in regard to environmental purchasing, highlighting some potential areas for involvement and improvement. The chapter begins with an overview of Walmart's approach, its cost orientation, and attempts to engage its suppliers globally and really measure impact. This is followed by a look at some other consumer product companies, including Procter & Gamble (P&G) and Method™ products. A common theme in consumer products is packaging redesign and reduction. This aspect of environmental purchasing is explored for each of the companies. Some of the initiatives that are unique to each company are also addressed.

The Walmart Initiative

When Walmart first announced its green initiative in 2006, the CEO at the time, Lee Scott, stated that Walmart wants to be, "… A good steward to the environment," and noted "As one of the largest companies in the world, with an expanding global presence, environmental problems are our problems."[1]

Walmart brought this same message to its suppliers in China in 2008 at its Beijing Summit, and reinforced this in 2012.[2] Because Walmart is not a manufacturer, it decided it could have the greatest impact by focusing on its transportation, its store sites, and of course, its suppliers. Perhaps it is that very public emphasis on using green supply chain methods that also reduce cost that has gotten Walmart so much publicity—as well as many followers—of its approach to green supply chain management.

Sustainability has become a very integrated part of everything that Walmart does. As noted by the world's largest sustainability research firm, Innovest Strategic Value Investors, "Sustainability is not a separate initiative at Walmart, as it is at many other firms. Sustainability is being integrated into all parts of the business."[3] Walmart expects a great deal from its suppliers. Walmart recognizes the huge impact that its supply base has, stating, "The footprint of Walmart's global supply chain is many times larger than its operational footprint and represents a more impactful opportunity to reduce emissions."[4] There are eight key practices that Walmart initially adopted to "green" its supply chain. Most of these had a significant and direct impact on Walmart's supply base. These practices are listed in Table 5.1. These approaches have evolved and been refined over time.

The Sustainability Consortium

In 2009, Walmart announced at a meeting of 1,500 suppliers that it was planning to launch a sustainability index that would provide a single source of information for evaluating product sustainability.[5,6] The goal

Table 5.1. Walmart's Key Initial Green Supply Chain Practices[7]

• Identifying goals, metrics, and new technologies
• Certifying environmentally sustainable products
• Providing network partner assistance to suppliers
• Committing to larger volumes of environmentally sustainable products
• Cutting out the middleman
• Restructuring the buyer role
• Consolidating direct suppliers
• Licensing environmental innovations

of the index was "to drive higher quality, lower costs and measure sustainability of products for the first time." To support this effort, and also in recognition of the fact that suppliers were receiving competing sustainability demands from many sources, Walmart formed The Sustainability Consortium to collaborate and provide input from suppliers, retailers, NGOs, universities, and the government.[8]

Today, The Sustainability Consortium appears to have grown into an NGO of its own right, independent of Walmart. It has a strong product lifecycle perspective and appears to be going strong, gaining members on a regular basis. Its efforts are clustered by industry, to focus on the issues important to various industries. The areas on foci are listed on its website and can provide insight for companies regarding the largest areas of impact of various products over their product lifecycles. Its stated goal is "...developing a standardized framework for the communication of sustainability-related information throughout the product value chain. The framework, called the Sustainability Measurement & Reporting System (SMRS) serves as a common, global platform for companies to measure and report on product sustainability. It enables rigorous product level Life Cycle Assessments to be done at a fraction of today's time and cost, and provides a platform for sustainability-related data sharing across the supply chain."[9]

The Sustainability Consortium has adopted a model known as Open IO, described as "... a fully accessible, transparent economic input-output life cycle assessment database that permits complete access to the user desiring a greater knowledge of the development of the database, and allows for model data and results to be available for other applications."[10] This model was created by Sylvatica, and is IP of New Earth, which is a not-for-profit developing and providing tools for sustainable development (newearth.info). The Sustainability Consortium staff keeps this model updated and available.

The model itself provides a great deal of insight into the lifecycle impact of many products over their lifecycles. In using the model, one can first select from a wide range of industries including apparels and textiles, education, electronics, lighting, and transport. Once that category is chosen, one can choose from a wide range of products within that industry. For example, apparel and textiles includes everything from accessories, carpets and rugs, and footwear to women's apparel. The electronics industry includes audio and video equipment, electronic computer manufacturing, biomedical equipment, and watches and clocks. Once a specific product is selected, in this case electronic computer manufacturing, the user can select from five various categories of damage for that product over its lifecycle. There are five broad categories of damage that can be chosen for analysis: climate change, which looks primarily at GHG implications; ecosystem quality, considering disappearance of species; human health impact in terms of adjusted life expectancy; resources in terms of use of nonrenewable energy; and water consumption. There is more detail on how each of these impacts is calculated on the Open IO website.

The model looks at the lifecycle stages of materials and services, transportation of inputs, energy inputs, direct operations, distribution, and consumer use. Looking at the "damage" category of climate change for electronic computer manufacturing, the two greatest lifecycle areas of damage are distribution (over 13% attributed to warehousing and retailing) and consumer use (over 58%, attributed to energy use). A primary suggested use of this model is to "… combine the information here with purchasing data to prioritize sustainability initiatives."[11] Thus, companies can use this model to understand where the greatest environmental impact of a product is over its lifecycle and where the greatest opportunities for improvement lie. For example, in purchasing computers, the amount of energy that the computer actually uses is a key contributor to greenhouse gas emissions. This tool can be an important part of communicating with suppliers on the importance of various sustainability initiatives. This also indicates an opportunity for companies to think about the way that they use computers and improve their energy use, perhaps turning them off at the end of the day, or having them revert to a power save mode after a certain amount of inactivity.

Walmart Carbon/GHG Focus

In 2010, Walmart formally asked its suppliers to look at the carbon lifecycle of their products from cradle to grave.[12] An initiative to reduce GHG was developed, which has three major parts:

- *Selection:* Walmart will focus on the product categories with the highest embedded carbon. This is defined as the amount of lifecycle GHG emissions per unit multiplied by the amount the company sells. To find the embedded carbon, the ASC reviewed the GHG emissions associated with all Walmart product categories. This approach ensures the project team focuses on the categories that have the greatest opportunity for reductions. Reductions can come from any part of a product's lifecycle.

- *Action:* For a project to be included as part of this goal, it must reduce GHGs from a product in either the sourcing of raw materials, manufacturing, transportation, customer use, or end-of-life disposal. Walmart must demonstrate it had a direct influence on the reduction and show how that reduction would not have occurred without Walmart's participation.

- *Assessment:* Suppliers and Walmart will jointly account for the reductions. ClearCarbon will perform a quality assurance review of those claims to ensure methodology, completeness, and calculations are correct. When the claims meet the quality assurance check, PricewaterhouseCoopers (PWC) will assess under consulting standards whether the defined procedures were followed consistently to quantify the reduction claim.[13]

In addition, Walmart made a commitment that by the year 2017, it will source 70% of its products from companies that use the Sustainability Index to evaluate their products and communicate the results to Walmart. In 2013, Walmart began using the sustainability index information to design its private label products.[14]

In many ways, Walmart is really a special case because of its huge global presence, power, and sheer sales volume. However, Walmart has used its power to positively influence the practices of consumer products and durables suppliers all over the world and has engaged other retailers, NGOs, and universities in the process. While few organizations can influence as much change in supply chain practices as Walmart, there is something that everyone can take away from the many supplier-oriented efforts that Walmart has implemented.

Procter & Gamble and Packaging

Procter & Gamble is a significant supplier to Walmart, but also engages in sustainability efforts in its own right. As a consumer products company, product design and packaging are a very important part of P&G's brand image. Packaging helps convey an advertising message, provides important information on product usage and safety, and also protects the product. In addition, the packaging adds significant weight and volume to the product, increasing product size, affecting space utilization and potentially increasing costs in transportation, distribution, and storage space at the store level. P&G recognizes the impact of its utilization of packaging. Sustainable packaging is one of its five sustainability themes. Procter & Gamble has a stated goal of "using 100% renewable recycled materials for all of its products and packaging."[15] The challenge for a company like P&G is to retain all of the value-added by packaging, using sustainable, recycled, and recyclable materials that use less space and weigh less. P&G has numerous examples of how it has done that. These efforts are often complementary with other P&G sustainability initiatives, such as product compaction. Its approach to packaging sustainability is listed in Table 5.2.

For example, in 2007, P&G introduced more concentrated versions of all its North American brands of laundry detergent. It assesses the impact of products across their lifecycle. Scope 1 "… sustainability benefits per unit of production include a 20% reduction in CO_2 emissions, 15% less solid waste production, 20% reduction in energy consumption, and 15% reduction in water consumption." By using and purchasing less energy for its processes, Scope 2 emissions (impact of generation of purchased power) is also reduced. Scope 3 benefits include millions of dollars saved in transportation, warehousing, end-of-life packaging disposal or recycling, and store shelf space. At the same time, it has reformulated all of its products to work well in cold water, which reduces consumer energy use. Further, Walmart noted that based on three years' sales of concentrated laundry detergent, it helped the environment through:

- Saving over 95 million pounds of plastic resin;
- Preserving more than 400 million gallons of water;
- Conserving over 520,000 gallons of diesel gasoline; and
- Reducing more than 125 million pounds of cardboard.[16]

Table 5.2. P&G's Five R's of Packaging Sustainability[17]

Reduce	Less material used, for example, through cutting the weight of existing bottles, detergent compaction, or the optimization of the amount of energy and raw materials result in lower cost and a reduced load on the environment.
Recycle	An increased rate of recycling reduces the need for new material and again reduces the environmental load.
Reuse	Recovering packages after they have been used, or allowing them to be reused through refills, reduces costs and cuts down on the need for raw materials.
Replace	Switching to alternative resources such as postconsumer recycled materials, which can reduce material requirements as well as carbon dioxide emissions.
Remove	Limiting or avoiding the use of certain materials from packaging where that can improve the safety profile, the environmental quality, societal acceptability, or the compatibility with waste management systems.

Walmart required all of its suppliers to transition to concentrated liquid detergent. P&G, which had tried concentrating its detergent in the United States earlier but pulled back due to poor consumer response, noted that it was important to have Walmart's support in this effort. If P&G was the only one selling smaller container sizes, an uninformed consumer would think that the P&G brand provided less value, due to the perception of getting "less product" for the money. More recently, P&G announced that it was also compacting all of its powdered detergents, which helps both the environment and reduces costs, as summarized in Table 5.3.

P&G takes a lifecycle perspective on sustainability, as can be seen in the detergent example. In its role as a supplier to supermarkets, other retailers, and ultimately the end customer, P&G's changes are reducing Scope 1, 2, and 3 emissions. Thus, P&G is affecting its supply chain and being a good environmental steward. All of these changes also affect P&G's suppliers. Thus, P&G must work closely with its packaging and raw materials suppliers on improving its formulations so that they are environmentally responsible. As P&G continues to focus on the utilization of fewer materials, P&G has to work closely with its packaging suppliers as its product size and dimensions are constantly changing. P&G takes a lifecycle approach to recruiting and retaining its suppliers, evaluating not only performance and price, but also carbon footprint, energy, and water usage as part of its supplier scorecard.[18]

Table 5.3. Benefits of Detergent Compaction[19]

Less energy in manufacturing
Fewer raw materials (meaning less is needed per dose)
Less product (and so less waste)
Less packaging (less material waste and they are easier to transport and lead to less emissions in the transport process)
Less space in warehouses (reducing storage, heating, and lighting impact)
Fewer chemicals, such as sulfate, being washed into surface waters

Method™ Products

The companies discussed in this book are all well-established companies who had to change the way that they did business in order to embrace the concept of sustainability and environmental purchasing. This can be a very challenging change for a corporate culture to embrace. However, there are companies that were founded with sustainability in mind. One such company is Method™ Products, a producer of nontoxic home care and personal care products. Method™ Products considers itself to be a "true green" company, meaning that it considers the impact it has on the environment in all things that it does, using only naturally derived, biodegradable ingredients. Rather than taking a "lifecycle" perspective, or a cradle-to-grave perspective, Method™ has adopted a cradle-to-cradle (c2c) approach. The "Cradle to Cradle®" concept is a material use cycle that seeks to eliminate waste and/or virgin resource extraction through the creation of closed/continuous loops. It embraces designing products for their lifecycle and avoiding as much resource use and disposal as possible at every stage of the product's lifecycle. Cradle to Cradle® traces a material from the time it is extracted to the point at which it is recycled/reclaimed."[20]

In addition, Method™ is one of over 600 Certified B Corporations in 24 countries and over 60 industries. Method™ is also a charter member of Certified B Corporations. The B Corporations are certified by the nonprofit B Lab to meet rigorous standards of social and environmental performance, accountability, and transparency. The overall philosophy is to provide a better, more sustainable way to do business that is best for the world.[21]

GREENSKEEPING
HOW WE DEFINE SUSTAINABILITY

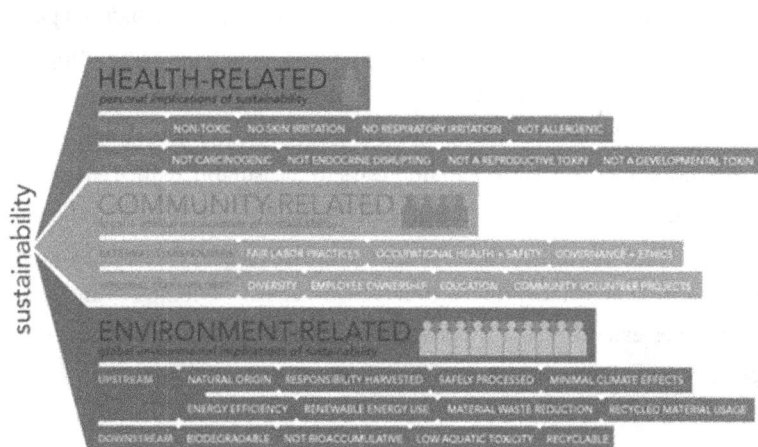

Figure 5.1. How Method™ defines sustainability.[22]

Figure 5.1 illustrates how Method™ defines sustainability. Its definition has many implications for supplier selection and supplier management. Virtually all of the items listed under environment-related concerns affect the supplier selection process for Method™.

Method™'s "greenskeeping" philosophy includes creating safe, effective, nontoxic products, using the most sustainable manufacturing and logistics processes, providing suppliers incentives for being more green, and pursuing certifications that support the environment and make Method™ a more sustainable company. In fact, it has 60 products that are Cradle to Cradle® certified and are the first cleaning products to receive that certification.

Because of the rigorous standards that it has for all of its ingredients and packaging materials, Method™ enforces very rigorous standards on its suppliers. All of its ingredients are independently assessed for health and environmental safety by the Environmental Protection Encouragement Agency (EPEA), an environmental research institute. Clearly, it is imperative that Method™ work very closely with suppliers to ensure that they adhere to its high-quality standards and nontoxic formulations, and that the manufacturers use environmentally responsible manufacturing processes. In addition, Method™ works with suppliers to eliminate all types of waste—energy, water, and materials. Its packaging is made primarily from recycled plastic and is designed for recyclability. It conducts research on the most recycled plastics and trends in recycling and designs its packaging following these trends. It must select and work with suppliers who are following those same trends. Many of Method™'s containers are refillable.[23] The ultimate goal is to completely avoid material waste.

In addition, if suppliers can find ways to measurably reduce energy use and the corresponding carbon it creates, Method™ contributes to the purchase of these solutions. These have included items such as heating/cooling fans, low-energy factory lighting, and solar panels.[24] Method™ and other companies extend their EP practices to environmentally conscious transportation. This is the topic of Chapter 6.

Summary

- There are many ways that organizations work with their suppliers as part of their EP approach, in order to reduce the environmental footprint of their supply chain.
- Walmart has had a tremendous impact on supply chain environmental sustainability practices globally due to its power, influence, and cost savings approach.
- There are an increasing number of organizations that help companies understand their relative environmental impact. Some, such as The Sustainability Consortium, have a lifecycle perspective.
- P&G focuses on a sustainability lifecycle perspective, giving attention to packaging, and creating products that are more environmentally friendly in consumer use.
- Most organizations have to make significant changes in how they do business to become better environmental citizens.
- Some organizations, such as Method™, were founded on the idea that the environment is a key part of their philosophy, and this is embodied in their purchasing and other supply chain decisions.

CHAPTER 6

What Is the Relationship Between Transportation and Green Sourcing?

Introduction

This chapter discusses the relationship of transportation to environmental purchasing. The relationship may be a very direct one. Purchasing often has input or even control of the transportation decision for inbound materials. In addition, purchasing affects the transportation footprint by determining which suppliers are selected, and thereby the location from which the required items are shipped. Purchasing also influences the transportation mode by determining the order size and how quickly the order is needed. All of these issues have a direct impact on the firm's carbon and environmental footprint. This chapter begins with a discussion of the environmental impact of transportation in general, and freight transportation in particular. Next, the chapter explores the role of purchasing in the transportation decision. The costs, speed, and environmental impacts of various modes are presented, and the SmartWay organization is introduced. Finally, an overview of the responsibilities and liabilities associated with transporting potentially hazardous material is discussed.

Transportation Impact

Transportation has a large impact on both the business and natural environment. From a business standpoint, transportation is the largest element of logistics costs; It constituted 63% of the total amount spent on logistics in the United States and amounted to 5.3% of the U.S. GDP in 2011— $806 billion.[1] Transportation is an essential element of every physical supply chain and critically affects an organization's customer service levels. It is a key to supporting international trade. Not surprisingly, it has the greatest negative impact on the natural environment of any industry other than power generation.[2] As a prime consumer of fossil fuels, generator of noise, and emitter of toxic chemicals, transportation is a large contributor to supply chains' carbon footprints.[3] While it is challenging to find specific, current data on transportation's impact on the environment, available data indicate that transportation uses about 26% of the world's energy. Within the broad area of transport, freight uses about 30% of that energy.[4] On a high level, there is a great deal of concern about the impact of freight transportation on the environment.

The Transportation Mode Selection Decision

Transportation modes vary in terms of cost, reliability (on-time), capability (ability to handle a wide variety of cargo), accessibility (to various locations), and environmental impact (GHG emissions). A comparison of the modes is shown below in Table 6.1.

It is interesting to note that there is essentially an inverse relationship between transportation speed and GHG emissions. Unfortunately, with so many long supply chains and customer demands for fast, reliable service, the worst polluting modes offer the most attractive options in terms of speed and access to the most locations as well as the highest costs per ton-mile.

Table 6.1. Ranked Comparison of Characteristics of Various Transportation Modes*

Selection determinants	Railroad	Motor	Water	Air	Pipeline**
Cost (ton-mile)	3	4	2	5	1
Transit time	3	2	4	1	5
Reliability	2	1	4	3	—
Capability	1	2	4	3	5
Accessibility	2	1	4	3	5
Security	3	2	4	1	—
GHG	2	4	3	5	1

*Ranking is based on 1 as the best (fastest, cheapest, lowest emissions) and 5 as the worst.
**Pipeline considers the impact of transmission. While reliability and security tend to be very high, they can also be sabotaged or siphoned in some parts of the world. Thus, their performance tends to be excellent or poor, with little middle ground.

Table 6.2. Summary of Fuel Efficiency by Freight Mode[5]

Mode	Ton-miles/gallon
Barge	533
U.S. Railroads	413
Truck	155

There is such a wide range of equipment in use in each type of transportation mode that it is difficult to give hard and fast comparisons among their efficiency and emissions. There are significant differences based on the age of the vehicle, maintenance, driver training, and other factors.[6] Table 6.2 provides some general comparisons that provide a good relative order of magnitude of freight efficiency.

Lower fuel efficiency translates into greater emissions, in general. But there are several different types of emissions that are associated with health damage. Carbon dioxide, also known as CO_2 or greenhouse gas, particulate matter (PM), and nitrogen oxide (NO) are the most commonly tracked and generally considered the most harmful to health in their current levels. Effects include respiratory problems, impaired fetal development, cardiovascular disease, and eye irritation.[7] Figure 6.1 shows the air emission per ton-mile by various modes. These data are also summarized in Table 6.3.

Based on the data included here, it is not surprising that the Natural Resource Defense Council (NRDC) advises companies to "avoid shipping by air" at all costs. However, in 2010 nearly 43.3 million tons of freight were transported by air worldwide, up from 30.4 million in 2000, which accounts for nearly 40% of all goods by value.[8] The NRDC also advises shippers to investigate individual carriers, as performance can vary immensely depending on their equipment and practices, reducing CO_2 by as much as 53% by using more fuel-efficient ships.[9] ShippingEfficiency.org provides a tool to calculate and compare efficiency by different types of vessels.[10]

The point of presenting this data is to reinforce the huge impact that transportation choices have on the environment, at both the modal level and the carrier level. Freight emissions have been frequently overlooked as a selection criterion for transportation mode, probably at least in part because it is not part of an organization's Scope 1 emissions. However, it

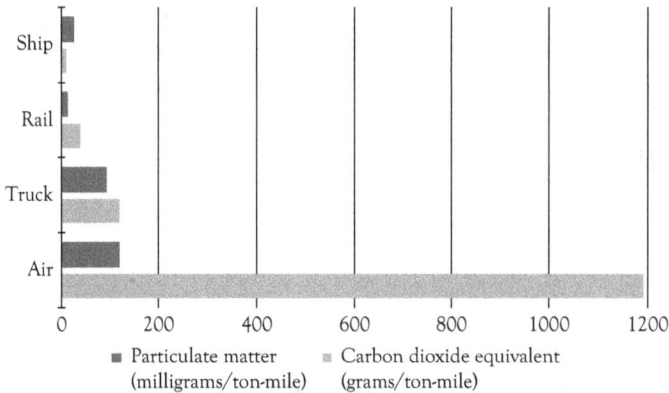

Figure 6.1. Emissions per ton-mile by freight mode.[11]

Table 6.3. Air Emissions per Ton-Mile[12]

	Ship	Barge	Rail	Truck	Air
Grams of CO_2e/ton-mile	11	40	40	119	1,193
Milligrams of NO_x/ton-mile	158	203	367	318	3,944
Milligrams of PM_{10}/ton-mile	25	59	13	92	119

is part of their Scope 3 emissions, which encompass the emissions of the supply chain. Thus, supplier location and how distance, access, and infrastructure affect the transportation choices and environmental footprint of the supply chain should be considered as part of EP.

To help companies examine the impact of their supplier locations on freight emission, the Clean Cargo working group, which is part of Business for Social Responsibility (BSR) has developed a carbon calculator to help assess the impact of various global transportation choices.[13] This is one example of the reality that freight transportation emissions have recently been receiving more attention. Some of this has been attributed to the fact that the less polluting modes are also less expensive per ton-mile. As companies have become more cost-conscious with slow global growth, transportation needs have been more carefully reviewed.[14]

Within the United States, the use of rail and intermodal shipping, combining rail with truck, has grown tremendously.[15] Domestic intermodal volume was up about 5.4% annually in a recent study. Continued growth in intermodal shipping combining rail and trucks is expected.[16]

Much of the cargo moved this way is sent in special containers or flatbeds that can be moved as a unit between trucks and rail, rather than loading and unloading cargo. While rail alone and intermodal may be slower than straight truck delivery, as shown in Table 6.1, using rail is cheaper and better for the environment. Using slower modes requires better demand data and more careful planning to achieve the same results, but it can be done. Dell has done this with its very customer-sensitive shipping of finished products. It is transitioning from shipping 80% by air and 20% by water, to about 70% by water and 30% by air.[17]

Walmart recognizes the importance of efficient transportation utilization to costs, customer service, and its environmental footprint. Like many other companies, Walmart had outsourced a considerable amount of its transportation, or utilized third-party carriers rather than expanding its own fleet over the past several decades. In the past several years, Walmart has decided that it wants to control more of its own outbound and inbound transportation. Walmart is said to have one of the largest green fleets in the United States. It has worked on improving its equipment efficiency and aerodynamics (with the Rocky Mountain Institute), load utilization, reducing weight, and improving routing so that in 2013, it is 69% more efficient than its 2005 baseline.[18] Today it is working with its suppliers on taking over deliveries from the supplier to Walmart distribution centers.[19] A Walmart spokesperson noted, that with Walmart taking over transportation for its suppliers, "It has allowed our suppliers to focus on what they do best, manufacturing products for us."[20] This strategy fits with Walmart's low-cost strategy as well as its environmental strategy, and recognizes the impact that inbound shipping from suppliers can have on both.

Empty backhaul, which involves moving a tractor back to its home base without a trailer, or with an empty trailer, is very inefficient and bad for emissions. These empty backhauls are often referred to as empty miles. One of the reasons that Walmart changed its transportation strategy to include picking up shipments from customers in its own trucks is to improve its backhaul utilization, which is good for the environment as well as for costs. But the customer shipment sizes and locations may not match Walmart's backhaul capacity in many cases. To utilize this space, Walmart teams up with companies such as Coyote Logistics. Coyote has created a business model to match unused trucks with shippers who need the capacity,[21] so that if Walmart cannot fill the truck itself, Coyote can help find a shipper that can use the space without interfering with Walmart's schedule.

Empty containers are also a serious environmental problem and very expensive. One shipping line alone, Maersk, indicates that it spends nearly $1 billion related to the shipping of more than 4 million empty dry and reefer containers back to where customers need them.[22] In 2010 as the economy rebounded, a shortage of shipping containers caught many carriers off guard. Container production for the past few years had been minimal, and many older containers reached the end of their useful life. The result for the carriers was very aggressive action to ensure availability and utilization of existing container fleets.[23] This had significant cost and environmental impact. Carriers evacuated empty containers from regions with surplus container capacity by deploying extra vessels just to transport the empties. They also reduced the length of time that empty containers were idle and delayed disposal of older containers.

The Transportation Selection Decision

The classic approach to supplier selection for transportation has been to first determine what mode is required than to select the specific supplier. When companies buy materials for use in production, they may be in a situation where they also have to arrange for delivery of those inputs, or where the supplier selects the mode. As a buyer, if you are focusing on EP, the transportation mode and carrier choice are important and should be weighed into the total supplier selection decision.

If your organization is a relatively significant customer, the control of transportation aspect may be a point of negotiation. For example, in the automotive industry, it is not unusual to require that your suppliers locate near your manufacturing facilities to support just-in-time (JIT) production using efficient "milkrun" transportation routing. The largest SmartCar factory in Hamburg, France, keeps it key suppliers on site, with their manufacturing facilities close by and coordinates delivery of parts exactly when needed. One such supplier is the door supplier. The right color doors, for the right model of car, have to be on site and available where SmartCar is manufactured. These short distance deliveries help to save on transportation and environmental costs.

In some cases, a company may have a separate logistics group that handles all of the outbound transportation of finished goods, because the transportation is for inbound materials, it is likely that purchasing will at least be involved in the transportation selection decision. To get an understanding of how important the environmental impact of inbound transportation is to a particular product or industry, see the Sustainability Consortium IO model.[24] When selecting suppliers, it makes sense to ask about where they purchase their materials and the method used to ship to them, the primary modes of transportation they use and whether they require their carriers or internal transportation group to have any environmental certification or affiliations.

An example of a company that considers the impact of the transportation of its materials and finished goods is Method™ Products, a manufacturer of "green" personal and home cleaning products. Method™ Products is a shipper member of SmartWay, a public–private partnership that focuses on improving transportation efficiency and reducing the associated emissions. SmartWay is presented in more depth below. As Method™ notes on its website, "We participate in the U.S. EPA's SmartWay program and partner with shipping companies that also participate, meaning that we are committed to reducing shipping energy use. We do this by strategic route planning, by effectively choosing where to operate our distribution centers, by using rail to deliver as many of our shipments as possible and by working with shipping companies committed to efficient freight technologies."[25] While Method™ does not, and never has, owned and operated its own fleet, reducing transportation emissions is so important to its total carbon footprint and its green philosophy that it has actually worked with SmartWay and J.B. Hunt, a trucking company, to implement biodiesel in its shipping. It has invested in the purchase of biodiesel trucks. As Method™ notes, "Working with one of our shipping partners, JB Hunt, we have built a fleet of super-efficient delivery trucks that run on minimum 20% biodiesel. This fleet emits roughly 30% less carbon per ton-mile than conventional trucks and is responsible for the delivery of our shipments in

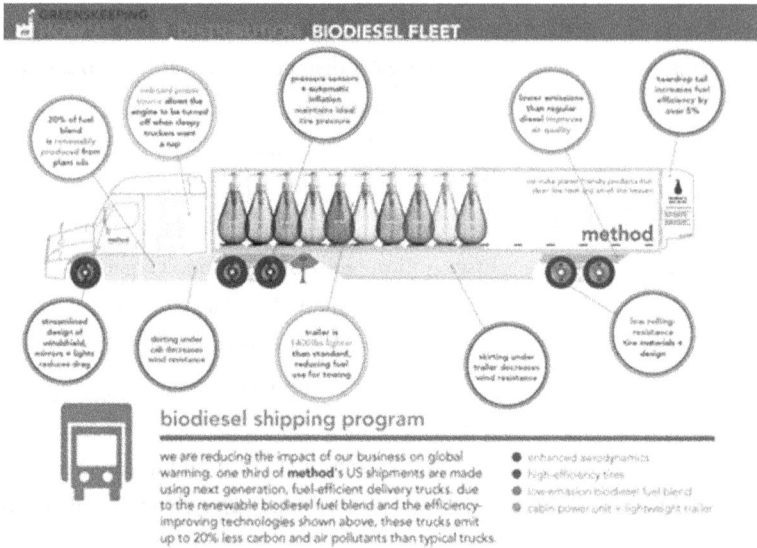

Figure 6.2. Emissions reduction and fuel saving features of Method™'s biodiesel fleet.[26]

California and the Northeast."[27] Some of the features that it incorporates into its biodiesel fleet are shown in Figure 6.2.

Environmental sustainability is integrated systematically into everything that Method™ does. The company was envisioned and developed as a green company, so it truly considers how a decision in one aspect of its operations affects the others, from cradle-to-cradle.

The routing of shipments and ports selected can also be important to reducing the organization's emissions and improving efficiency. Anywhere that the organization's demand patterns have changed, it is worth reviewing if there are significant bottlenecks in terms of increasing efficiency. For example, Patagonia, a hip and environmentally conscious clothing company, outgrew its distribution center near Los Angeles, and moved its primary distribution from Asia to Reno instead. Yet, it kept bringing its shipments in through the Port of Los Angeles/Long Beach, which has a reputation for being one of the most congested ports in the United States. Patagonia hired a consultant to review the routing. It recommended changing its receiving port to Oakland. As a result, Patagonia not only saved $324,000 in shipping costs (transloading and transportation mileage) per year, but reduced its inbound supply chain carbon by 31%. Most of this came from reduced truck mileage: 523 miles from Los Angeles/ Long beach to Reno, versus 229 miles from Oakland to Reno. Patagonia concluded that this change was long overdue.[28]

Transportation-Oriented Environmental Groups

There are several environmental organizations that focus on reducing the environmental impact of freight transportation. One is the Clean Cargo Working Group of BSR, a not-for-profit organization that focuses on the "business of a better world."[29] The Clean Cargo Working Group is part of the BSR that focuses specifically on global transport and provides assessment of many shipping lanes and carriers. The Clean Cargo Working Group describes itself as "a global business-to-business initiative made up of leading cargo carriers and their customers, dedicated to environmental performance improvement in marine container transport through measurement, evaluation, and reporting."[30] Members benefit through participating in the creation and use of tools related to measuring, evaluating, and reporting the environmental impacts of global goods transportation. It provides a standardized method that carriers can use and shippers can understand to communicate environmental impact and focuses on how to reduce such impact. Thus, this can provide very useful environmental impact information in the ocean carrier selection decision, keeping in mind that emissions among carriers can vary dramatically, depending on equipment and speed. Its members carry 60% of the global freight.

Another NGO that focuses domestically in the United States is the Rocky Mountain Institute (RMI), whose mission is "to drive the efficient and restorative use of resources."[31] That includes a group that works specifically on transportation, including trucking. "The RMI's vision for the trucking sector's next decade of development focuses on a radical shift to more cargo being moved with less fuel." It works closely with the North American Council for Freight Efficiency (NAFCE), which was formed out of a RMI freight transport meeting. The RMI's focus on freight transportation is reduced fuel consumption through the following:

- Increasing the aerodynamics of new trucks and trailers
- Improving older trucks through retrofits

- Increasing the use of tires with low rolling resistance
- Decreasing the weight of new trucks and trailers to increase the amount of cargo delivered per weighed-out load
- Monetizing efficiency and productivity gains, thus prompting uniform industry incentives for driver training, load and route optimization, and so on
- Increasing freight hauled per trip by mandating aggressive and internationally competitive fuel economy, length, and weight standards
- Accelerating the adoption of efficient technologies by developing supportive policies and funding mechanisms
- Integrated vehicle (and fleet) level design that utilizes the synergies of efficiency in one subsystem to enhance the efficiency of others
- Scaling the viability and production of sustainable energy-dense fuels and hybrid power train technologies to complement end-use efficiency improvements[32]

A trucking company that is an active member of NACFE is probably environmentally proactive. This could be used as one indicator for companies in selecting trucking firms.

Walmart is one of the companies that has worked with and utilized the RMI's expertise in revamping and dramatically improving its freight efficiency.

Probably the best current external indicator of a company's environmental transportation efforts today is membership in SmartWay. SmartWay began in 2004 and is a voluntary public–private partnership under the auspices of the U.S. Environmental Protection Agency. SmartWay is "a public/private collaboration between the U.S. EPA and the freight transportation industry that helps freight shippers, carriers, and logistics companies improve fuel-efficiency and save money, and in the process reduce emissions and improve air quality."[33] SmartWay has members that are freight shippers, truck carriers, rail carriers, logistics companies, and multimodal carriers. Carrier members agree to track their fuel consumption and freight carriage, and improve their performance each year.

In addition to providing tools for tracking operations, SmartWay provides low-cost financing for efficiency improving equipment and upgrades,

Table 6.4. SmartWay Emission Metrics

• Grams/mile of carbon dioxide (CO_2)	• Grams/ton-mile of carbon dioxide (CO_2)
• Grams/mile of oxides of nitrogen (NO_x)	• Grams/ton-mile of oxides of nitrogen (NO_x)
• Grams/mile of particulate matter 2.5 microns (PM2.5)	• Grams/ton-mile of particulate matter 2.5 microns (PM2.5)
• Grams/mile of particulate matter 10 microns (PM10)	• Grams/ton-mile of particulate matter 10 microns (PM10)

performance rankings of carriers, information on energy-efficient equipment, and a list of all members and affiliates. This information can be used in selecting carriers. Buyers can also ask potential suppliers whether they are affiliated with SmartWay themselves and whether they require that a certain percentage of their carriers or shipments are transported via Smart-Way. The publicly available data is extensive and tracks member performance according to the metrics shown in Table 6.4.[34]

Companies who are not carriers can join SmartWay as shippers. They can track their own company's transportation footprint, if applicable, as well as the performance of their carriers, and how that affects their supply chain footprint. The goal is continuous improvement. An increasing number of companies, such as Starbucks, Chrysler, and Kroger, are strongly encouraging all of their carriers to become SmartWay members, so that these shippers can proactively reduce their Scope 3 emissions. Thus, SmartWay affiliation can be another tool for selecting suppliers in support of EP, whether looking specifically at buying transportation, or whether considering the transportation approach of a potential materials supplier.

Transporting Hazardous Materials

There are numerous laws and requirements associated with transporting hazardous and flammable materials. Anyone who is involved in purchasing such materials should become very familiar with the laws governing their transportation in order to avoid environmental damage, safety issues, and creation of both personal and company liability. Thus, the transportation of hazardous materials goes beyond EP concerns alone. "The Hazardous

Material Transportation Act (HMTA) was published in 1975. Its primary objective is to provide adequate protection against the risks to life and property inherent in the transportation of hazardous material in commerce by improving the regulatory and enforcement authority of the Secretary of Transportation."[35] Some organizations, such as IBM, have opted to eliminate all hazardous materials to the extent possible. Elimination of hazardous materials reduces supply management and supply chain complexity significantly.

Key Points

- Transportation has a tremendous negative impact on the environment, which is frequently not specifically incorporated into environmental purchasing and supplier selection.
- Supply management can work to reduce the environmental impact of its purchases, while keeping business objectives and the total cost and environmental footprint in mind.
- Different modal choices and even different carriers within a mode can have a vastly different environmental impact, which should be considered in the selection decision.
- Improved transportation management, including good vehicle maintenance, efficient routing, and reducing empty backhauls, are all significant opportunities for total cost reduction and emissions reductions.
- Numerous organizations have emerged to help both carriers and shippers better understand and reduce their environmental footprint from transportation. These include SmartWay, the Rocky Mountain Institute, and the North American Council for Freight Efficiency and the Clean Cargo Working Group.
- SM involvement in transportation is also important in terms of ensuring compliance to the many laws governing the movement of hazardous materials.

CHAPTER 7

A Path Forward

Introduction

The previous chapters have provided information about how various organizations view EP, different levels of environmental purchasing, from buying green to designing for environment, and some successes and best practices. With that information as background, this chapter will review and integrate some of the information in previous chapters to help provide a framework for how to incorporate EP in your organization. The chapter begins by providing deeper insights into how to start an EP process, key considerations and differences at various stages of implementation and how to institutionalize EP as part of your organization's practices. Whether or not your organization is engaged in EP, and no matter how long it has been engaged in EP, there is always room to reevaluate where you are and realign and improve your processes.

Where to Begin with, Implementation or Assessment of Current Practices?

Whether your organization is just beginning to think about EP and sustainability, or whether you are a "true green" company like Method™, founded on sustainability principles, the best place to begin reviewing your EP practices is by first understanding the organization's sustainability philosophy and objectives. There are a number of ways that you can do that. If you work for a large company, it may have a sustainability report that it publishes annually. These reports, often called Corporate Social Responsibility reports or Environmental Reports, detail the organization's philosophy toward environmental sustainability and often toward social responsibility. Although there is no standard required format, an increasing number of organizations are using the Global Reporting Initiative (GRI) guidelines, so that there is growing consistency in the types of items reported around emissions, water usage, and much more.

Motorola's environmental mission states "Protecting the Natural Environment is Important to our Customer's and our Business." For example, it works on reducing electronics waste while still making the customer's life easier. It designs products to lessen energy use and also to use compatible and universal chargers. At all of its worldwide facilities, rigorous EHS policies and systems are in place. It has specific and aggressive goals to reduce emissions, water use, and waste. Its facilities are certified to ISO 140001 standards and OHSAS 18001.[1]

Target is an organization that states "From the way we build our stores to the products on our shelves, environmental sustainability is integrated throughout our business." Target's website has a dedicated section for responsible sourcing and focuses on sustainability of its supply chain. It believes that the future of its business depends on taking care of the resources that are here today. Target works with industry partners and experts on ways to reduce the environmental impact of the products it sells, from their packaging to the transportation that carries them to the stores. Working with its supply chain partners helps to create better tools and more efficient processes for producing and shipping the products over time and to use what it learns to improve its own practices. In 2011, they became a founding member of the "Sustainable Apparel Coalition" focused on reducing environmental and social impacts of apparel and footwear products around the world. This coalition created a measurement tool called "The Higg Index," which is a tool to help the industry measure the impact of materials, packaging, manufacturing processes, and transportation on the environment.[2]

Motorola and Target are just two organizations with mission statements around environmental sustainability. Many organizations have such statements, and many publish an annual report on sustainability. CorporateRegister.com maintains a database of published reports and claims to have over 46,000 reports across more than 9,700 companies.[3] Looking for these statements and published reports will provide insight into a given organization's sustainability emphasis and tracked metrics. Reading and reviewing your organization's sustainability report can also provide insight into areas that you might want to emphasize. A company's annual report to shareholders and other public statements might also provide insight. This can provide guidance as to where purchasing should be focusing its efforts when working with suppliers. For example, in reading its annual report, consumer goods company P&G is clearly focusing on the areas of packaging and production compaction.[4]

If your organization does not have any formal written statements, as is true of some smaller companies and companies that have not been focusing on sustainability, a good place to begin is to find out who is in charge of sustainability in the organization. If it is not clear, or it is not a formal role, then it is often the same person who is in charge of employee health and safety. You may have to ask around. If it appears that there is no formal approach in the organization, this could be an opportunity to propose one, at least in the purchasing area. It is very important that whatever you propose is in line with the organization's overall objectives. For example, a small distribution organization that is trying to grow regionally and does not have a formal approach to EP or sustainability in general is probably not going to begin by buying only renewable energy sources. But it may begin by buying more recycled products for internal use, looking at the packing it uses to get products to its customers, or perhaps even offering some more environmentally friendly products for sale.

The key is that whatever you do, or propose, it has to fit with and support the company's goals. Companies like Method™ are actually willing to pay more for sustainable inputs because it is fundamentally a part of what the company stands for. For example, Method™ invested in its own biodiesel trucks, which is definitely more expensive than regular diesel is today.[5] Most companies will be interested in being cost neutral in EP to the greatest extent possible—they do not want EP to cause profit margin pressure or force it to raise its prices. The perception of the added expense of EP is one of the greatest barriers to its effective implementation.

There are many cost-related benefits possible when implementing EP. Looking at ways to achieve cost reductions or value improvements simultaneously while implementing EP can be a very strong selling point. A key here is that any waste that can be taken out of the system is both good for the environment and also saves on costs. This approach to simultaneously reduce waste, cost, and its environmental footprint has certainly been effective for Walmart.

With the organization's environmental philosophy, business mission, and constraints in mind, it is possible to get a good idea of how broadly and deeply one might begin. Corporate culture and willingness to change also plays a role in how to position EP efforts. The potential benefits should be positioned favorably in line with the organization's goals, whether they are cost savings, image, or safety.

Buying green is a great starting point, and that may be as far as some organizations can go, for a variety of reasons. Limitations could include resource constraints, costs, and just the fact that a small or remotely located company may have limited choices of suppliers due to its size, or its distance and shipping costs and impact from major hubs.

Team Members

Unless you work for a very small company, others in your organization will be affected by and affect your EP approaches. It is good to form a team to develop the organization's EP approach, or to assess the current approach. The team should include purchasing and key stakeholders. These could be people who use what SM buys, the person with environmental responsibility in the company, and even accounting, to set up measurements and track cost or savings. It is helpful to have people who are strongly committed to improving the state of the natural environment. The passion that these team members bring can keep the energy of the team high and positive.[6] Many environmental ideas begin with the passion of a single employee wanting to make a change; there are many excellent initiatives that begin at the lower levels of the organization. Membership of the team will vary, but it is essential to have a good base of support and a good perspective of the implications of EP for the total company. This team will develop recommendations for the overall EP approach, help research sources, and help implement the overall approach among their peers.

Activities to Move EP Forward

One of the key issues that can really help move EP forward, as well as an organization's environmental approach in general, is the organization's philosophy toward waste reduction. As a starting point, one of the most impactful things that an organization can do to help the environment is to reduce waste. Whether or not a frugality mentality is already part of the organization's overall philosophy, it needs to be part of the EP philosophy because a very fundamental part of improving the environment is examining the way that things are used, and reducing waste at every part of the product lifecycle. People and organizations that understand this concept will be more successful in fully implementing the aspects of EP that are most relevant to their companies.

Industry Groups

Numerous examples of very effective industry groups, NGOs, and public–private partnerships have been presented throughout this book. There are many organizations that have extensive knowledge regarding how to improve the environment through environmental purchasing efforts of all types. The best advice for those new to EP is not to reinvent the wheel. Find an organization that has already done a lot of work in this area. Visit its website, and see how it fits with what you are trying to accomplish. If it seems beneficial, you may want to join the organization, attend meetings, and learn first hand from others who are at various stages of the journey to sustainability through EP. New approaches to EP are being implemented each day, so involvement in such organizations may be ongoing. As an example, the Sustainability Consortium's noncorporate members include organizations listed in Table 7.1. This list continues to grow as the potential impact of the Sustainability Consortium is realized.

Certifications

There are numerous environmental certifications. Some are applied broadly, like ISO 14001, while others, like the EICC, are more specific to the electronics industry. Unless you are a very large company, you probably cannot afford to do your own audits of suppliers, other than perhaps on an exception basis. One of the things that the EP team can do is investigate environmental certifications and see which might be more relevant

Table 7.1. The Sustainability Consortium's Government and NGO Members

Business for Social Responsibility
Caredefra
Environmental Defense Fund
United States Environmental Protection Agency
Forest Sustainability Council
Fundacion Chile
Sik Science Partner
Sustainable Forest Initiaitve
World Wildlife Federation
National Resource Defense Fund

to its suppliers. Also, talk to others in your industry or attend meetings for professional organizations like the Institute for Supply Management or the Council of Supply Chain Management Professionals. Different types of certifications may be relevant to different types of suppliers. If it is a widely held certification, like SmartWay in the trucking sector, you might decide to make this certification a requirement for doing business. Many companies will only buy Energy Star certified electronics or give strong preference to certified organic ingredients. If it is a newer certification or monitoring program, like The Sustainability Consortium, you might want to include it as criterion for selection, with the understanding that suppliers would have to join and participate in indexing within a couple of years. Certifications can be very helpful for meeting basic requirements and provide a good screening mechanism for potential suppliers. The certification should be relevant to the organization's requirements and meaningfully reflect supplier's positive environmental behavior. There are literally hundreds, if not thousands, of environmental NGOs and government organizations globally. Many of these partner with each other. With all of the environmental organizations in existence today, it requires some homework to determine which might be appropriate for your organization to join, and where you can learn the most that is relevant to your situation.

Segment Your Supplier Base

Not all suppliers are equal in terms of their importance and environmental impact. Thus, suppliers should be segmented in terms of their environmental impact on the organization, the organization's environmental footprint, and a number of other factors. If your organization purchases anything that is hazardous or regulated, the suppliers that you buy these items from must be very carefully screened. IBM and many other organizations take this approach. It has a high potential impact in terms of the damage that could be done. This is true even if the item is purchased in very small quantities. These items obviously pose a high potential risk if mishandled.

Things that are not hazardous could also have a high potential impact on the environment, due to the sheer amount of the item that is purchased. For example, New Belgium Brewing Company, the makers of Fat Tire and other ales, examined the environmental footprint of their supply chain, and found that refrigeration, the manufacturing of the glass bottles, and producing barley and hops are the top carbon contributors to its supply chain. While it struggles with how to refrigerate beer more efficiently, it is evaluating recycled glass bottles and organic barley and hops as ways to reduce its carbon footprint. These moves involve investigating different suppliers, or altering relationships with existing suppliers to meet its revised needs.[7] Thus, the items with the most impact should be examined.

On the other hand, the options available are also an issue. For example, New Belgium currently does not have another option for its refrigeration methods among commercial technologies. So it needs to consider how to improve the way that its current technology is used if it wants to have an impact.

Yet another related issue is the relative power or influence of the firm on its supply base. Earlier in the book, many examples were provided of Walmart and its significant influence on its supply base. Important customers can sometimes influence their suppliers. This is especially true if the suppliers believe that the customer is committed to doing business with them in the long run.[8] If the organization is a small or noninfluential player, it may have to choose from options already available in the market from existing suppliers.

Should Purchasing Lead the Way in Environmental Practices?

Some EP activities may be relatively easy to implement and require little support. These include activities like giving preference to items that are renewable or made from recycled materials that meet all of the organization's other requirements. If there are people in supply management who are passionate about EP and will serve as champions, then it may be a good place from which to lead the organization's environmental initiatives and influence other areas. An important issue in the success of any EP practice, and sustainability practice in general, is the willingness and ability to dedicate resources to its successful implementation and ongoing operations. This is why the idea of a passionate champion or group of champions can be so important. Those involved in sustainability may have to do research and get involved in professional associations that meet outside of work hours.

Supply management will be more successful as a leader in sustainability if it is already a well-respected area within the organization. Since supply management is already viewed as a key area within the organization for reducing costs and increasing value, it should position its efforts accordingly. If people can see that there are specific, measureable benefits in adopting EP, beyond "feel good" metrics of reducing CO_2, they are more likely to be supportive. It is good to position efforts to highlight as many real benefits as possible, as different benefits may appeal to different people. SM also has to be ready to address concerns, such as "recycled paper is more expensive, lower quality...," and so on.

Thus, supply management can lead through its actions, and through its meaningful results. As indicated earlier in the book, it is not enough to simply buy more sustainable items. The way in which items are used is also critical. Supply management can play an important role in helping the organization to focus on the areas within the product lifecycle that make the most difference to the items that the organization is purchasing, using, and making.

The Product Lifecycle Revisited

As review of the product lifecycle concept or the work of the Sustainability Consortium supports that environmental impacts vary tremendously at different stages of the product lifecycle. They also vary tremendously among different industries. It is worth taking the time to look at the Sustainability Consortium's calculator to understand what the impacts are for various products that your organization makes or uses.

In the trucking sector, there are many technologies available to improve a truck's footprint and increase its gas mileage. These include auxiliary power units, particularly solar ones, to heat and cool the cab when the truck is parked, rather than running the engine, aerodynamic design, alternative fuels, and improved engine design. However, the most impactful way to improve mileage and reduce the carbon footprint of an existing fleet is through driver training. By teaching drivers to reduce their idling time, moderate their speed, reduce rapid acceleration, and test and maintain their tire pressure, they can reduce fuel usage as much as 25%. At the same time, it can lengthen tire life, and reduce vehicle wear and tear and maintenance costs.[9] Thus, for companies with their own trucking fleet who are trying to understand the best way to invest in improving fleet performance and reducing their carbon footprint, driver training is the place to begin.

For a company that makes shoes, like Timberland, it is the choice of the raw material—the design—that really drives the carbon footprint. For example, its Winter Park Slip on Ski Boot has a carbon footprint of 121 pounds, 112.5 of which come from the raw material. The biggest impact comes from the leather, but the rubber sole and ethyl vinyl acetate (EVA) insole are also major sources of CO_2.[10] The materials are chosen as part of the design process. In many ways, design really impacts the overall carbon footprint of a product more than anything else. Of course, design dictates cost. However, it also dictates the materials, and the materials used will also dictate the way a product can best be used and its expected lifecycle. Product design strongly influences the energy consumed in product use, and the ease of recycling. The implications here are quite complex and vary significantly with the user. If lower-polluting inputs were chosen, would that increase sales? By using these relatively high-polluting, but durable inputs, does this actually increase the useful life of the product? But more importantly, does it increase the amount of time before the consumer will dispose of and replace the product?

Very few companies have the time and resources to really delve into a deep understanding of their products' footprints. The Sustainability

Consortium's Open IO provides a place to start in understanding the biggest impacts for numerous products. This can help companies better understand where to begin the process of improving their environmental footprint in general, and where to focus their EP efforts more specifically.[11]

It is important to keep in mind that while EP issues are critical, they are not the only concern of the organization, nor are they even always the most important concern. Patagonia has "the footprint chronicles" on its website to show anyone interested where it sources from, its supply chain flows, and why it makes the decisions it does regarding sourcing and other supply chain decisions.[12] The Talus jacket has a carbon footprint of 66 pounds, 47 of which comes from producing oil-based polyester. It considered using recycled materials, but they were not available where the product was made in Asia. The recycled material would have had to be found in the United States and shipped to Asia, at great expense. Patagonia believes that the price point would have killed the product. In this situation, the cost concern outweighed the environmental concern, and the product was made from virgin fiber in Asia.[13]

Supply management can play an important role in the design process by helping designers and others understand the importance of design decisions on the lifetime impact of a product. It can also support improved designs by reaching out to key suppliers and making them aware that the organization is interested in new, environmentally sustainable products, processes, materials, and solutions, so that suppliers will approach supply management with new ideas. Supply management can also help tremendously if it is a member of the new product development team by ensuring that lifecycle issues are addressed early in the idea conception stage.

Engaging the Supply Base

Suppliers are experts in what they do—or they should not be suppliers of that item. As purchasers we buy supplier's capabilities. As such, companies should consult with suppliers about materials and parts before they commit to designing such items into products. Such consultation may influence the type of relationship that you have with the supplier, making the supplier more important to your organization over time. Earlier in the book, it was noted that it is very important to work closely with suppliers and keep them abreast of changes you are making, or anticipate making, to reduce your product footprint. But the suppliers may also have knowledge of important changes in materials, technology, and application by working with different industries and companies. Those suppliers that are committed to the sustainability of the supply chain will more likely share good ideas on how to improve the footprint of the products.

Suppliers can be an important source of untapped innovation in improving design for the environment. One area where this is very true is packaging. With all of the emphasis on lighter weight, recycled and stronger packaging materials, leading packaging suppliers are very knowledgeable about trends and can suggest the best options to meet an organization's needs and budget.[14]

Measure the Right Things

In order to get good support and recognition for EP efforts and to motivate employees to focus on EP, good measurement is important, both internally, as well as externally. Many of the measures will be the same. The internal measures may be shown by business unit or project, in finer detail. The external measures, such as the items included in a sustainability report, will likely be shown in aggregate, except to highlight specific stories and successes. The measures should be tied to your organization's goals. Among the most common measures are cost savings and GHG reduction. However, depending on your goals, there can be numerous other internal measures, such as those shown in Table 7.2.

Table 7.2. EP Outcome Measures—Internal

• Cost savings
• GHG reduction
• Percentage of energy star items purchased (where energy star exists)
• Percentage of recycled products purchased
• Percentage of office waste to landfill vs. recycled
• Percentage of production waste to landfill vs. recycled
• Extended life (delayed replacement) of capital

There is also a framework called the Sustainability Measurement and Reporting System (SMRS), which serves as a common, global platform for companies to measure and report on product sustainability. This system, developed by The Sustainability Consortium, enables rigorous product level lifecycle assessments to be done very quickly and cost efficiently. This system also provides a platform to share sustainability-related data across the supply chain. Measurement is important in motivating behavior. Acknowledging and rewarding improvements is also very valuable. This can include everything from including a story on the organization's website or newsletter to cash rewards of all sizes.

The road to EP and sustainability in general is a journey, not a destination. Each step brings you closer to a final destination. However, sometimes these steps are incrementally small with many barriers and obstacles to overcome. Other times the steps are huge and require a leap of faith and commitment from those involved in helping to improve an organization's environmental footprint.

Key Points

- Every effective environmental purchasing initiative begins with developing an understanding of the overall organization's environmental sustainability objectives and developing approaches that support those objectives.
- Many organizations have formal environmental sustainability objectives that are published on their websites, in their voluntarily published sustainability reports, and even in their annual reports to shareholders.
- Implementing EP usually should involve a team of key stakeholders that will be affected by EP decisions.
- Membership in sustainability-oriented organizations such as the Sustainability Consortium and environmental certifications can be effective ways to move your organization's sustainability practices and focus ahead.
- When determining where to focus your sustainability efforts, segment your supply base according to environmental impact and then begin to analyze where you can make the most difference.

- The environmental impact of a product or service varies significantly over its product lifecycle and should be a consideration when determining where in the lifecycle and supply chain to focus EP efforts.
- Suppliers are selected based on their capabilities. Supply management can help leverage the capability of suppliers in sustainable initiatives, working with existing suppliers or new suppliers to incorporate environmental innovation into product and process changes.
- It is important that good measures of EP are developed to measure the progress and contribution of EP in order to sustain efforts and improve EP practices.
- The "green" or environmental sustainability movement is not a fad. Organizations see that greater environmental emphasis in their product and service offerings not only help their image, but also may reduce inefficiency, waste, and corresponding cost.

Notes

Chapter 1

1. EPA (2012).
2. EPA (2012).
3. Resourcities (2012).
4. Eurostat (2011).
5. Lee, O'Marah, and John (2012), p. 29.
6. U.S. Census Bureau, Survey of Manufactures (2013).
7. U.S. Census Bureau (2012).
8. Walmart (2008).
9. See definitions from PCF World forum (2012), and European Union (2012).
10. EPA (2012).
11. European Union (2012).
12. Honda (2013).
13. Lee et al. (2012), p. 29.

Chapter 2

1. EPA (2013).
2. Brown (2011).
3. Just Live Greener (2012).
4. Dutch Waste Management Association (2012).
5. Sunny Delight (2011); Iobst (2012).
6. Environment Agency Europe (2012).
7. Clean Energy Ideas (2012).
8. Sprint. Re:cycle (2013).
9. Starbucks (2012).
10. LeBlanc and Rick (2012).
11. CHEP (2013).
12. Business Dictionary (2012).
13. Kraft Foods (2012).
14. Zsidisin and Siferd (2001); Carter and Carter (1998); Carter and Narasimhan (1998).
15. Engel (2011).
16. Zsidisin and Siferd (2001).

17. EPA (2012). EPA (2013).
18. Adapted from NASPO (2013).
19. Adapted from Noharm.org (2013).
20. UNEP/SETAC Life Cycle Initiative (2013), p. 14.
21. UNEP/SETAC Life Cycle Initiative (2013), p. 14.
22. Hewlett-Packard (2012).
23. McDonough and Braungart (2002).
24. EPA (2012).
25. Carbon Neutral Company (2013).
26. Coca-Cola Company (2013).
27. Walmart (2012).
28. Environmental Leader (2008).
29. IBM (2013).

Chapter 3

1. Kraljic (1983).
2. Kraljic (1983).
3. Pagell, Wu, and Wasserman (2010).
4. Krause, Vachon, and Klassen (2009).
5. Pagell et al. (2010).
6. Kraljic (1983); Krause et al. (2009).
7. IBM (2002), p. 4.
8. Ellram and Tate (2003); IBM (2002).
9. IBM (2011).
10. Leavoy (2011).
11. Apple (2012).
12. Apple (2012).
13. EICC (2012).
14. Nokia (2013).
15. Jackson (2011).
16. Callahan and Tohomas (2009).
17. Bilodeau (2013).
18. Rogers and Tibben-Lembke (1999).

Chapter 4

1. Carson (1962).
2. U.K. National Measurement Office (2013).
3. Mersen (2013).

4. European Union (2013).

5. Smock (2007).

6. Tate, Ellram, and Dooley (2012).

7. EICC (2008).

8. Tate, Ellram, and Dooley (2012); EICC (2012).

9. EICC (2012), List of Members.

10. Dell (2006), p. 26.

11. Dell (2012).

12. Hewlett-Packard (2006), p. 64.

13. Hewlett-Packard (2011).

14. Hewlett-Packard (2011).

15. Hewlett-Packard (2006); Hewlett-Packard (2011).

16. Tate, Ellram, and Dooley (2012).

17. Tate, Ellram, and Dooley (2012).

18. Kanari, Pineau, and Shallari (2003).

19. Environment Agency (2013).

20. Gerrard and Kandlikar (2007).

21. Honda (2013), Product Recycling (3Rs).

22. Lexus (2013).

23. USCAR (2013).

24. USCAR (2013).

25. Ball (2009).

26. E-stewards (2013).

27. E-stewards (2013).

28. E-stewards (2013).

29. EPA (2013).

30. Pelley (2009).

31. E-stewards (2013); Pelley (2009).

32. E-stewards (2013); Pelley (2009).

33. EPA (2013).

34. EPEAT (2013).

35. EPEAT (2013).

Chapter 5

1. MSNBC (2013).

2. Duke (2012).

3. Nandagopal and Sankar (2009).

4. Walmart (2010).

5. Walmart (2009).

6. Cheeseman (2010).
7. Plambeck and Denend (2007).
8. Walmart (2009); Cheeseman (2010).
9. Carlson (2013).
10. Sustainability Consortium (2013).
11. Sustainability Consortium (2013).
12. Rosenbloom (2010).
13. Walmart (2010); Rosenbloom (2010).
14. Duke (2012); Rosenbloom (2010).
15. Procter & Gamble (2012).
16. Walmart (2007).
17. Procter & Gamble (2013).
18. Wasik (2010).
19. Procter & Gamble (2013).
20. Method (2013).
21. Certified B Corporation (2013).
22. Method (2013).
23. Method (2013).
24. Method (2013).

Chapter 6

1. Wilson (2012).
2. Himanen, Lee-Gosselin, and Perrels (2004); Wu and Dunn (1995).
3. Ellram and Golicic (2013).
4. International Energy Agency (2008).
5. American Waterways (2009); Davis, Diegel, and Boundy (2011).
6. For more insight into some of the issues that affect emissions and fuel efficiency of various modes, see NTL (April 2010).
7. For more information on specific health impacts see EPA (2013), Effects of Air Pollutants-Health Effects.
8. World Bank (2013).
9. NRDC.org Clean By Design: Transportation (2012); World Bank (2013).
10. ShippingEfficiency.org (2013).
11. NRDC.org Clean By Design: Transportation (2012); Texas Transportation Institute (2011) for Barge emissions, adapted to comparable units of measure as a percentage of truck emissions.
12. NRDC.org Clean By Design: Transportation (2012).
13. BSR (2012).

14. O'Reilley (June 2012).
15. O'Reilley (June 2012).
16. Wilson (2012).
17. Manners Bell (July 10, 2010).
18. Walmart (2013).
19. Straight (2010).
20. Wolf, Burritt, and Boyle. (2010).
21. Joyner (2012).
22. Churchill (2012).
23. World Shipping Council (2011).
24. The Sustainability Consortium (2013).
25. Method (2013).
26. Method (2013).
27. Method (2013).
28. Patagonia (2013).
29. BSR (2013).
30. BSR (2013).
31. Rocky Mountain Institute (2013).
32. Rocky Mountain Institute (2013).
33. EPA (2013).
34. EPA (2013).
35. EPA (2013).

Chapter 7

1. Motorola and the Environment (2013).
2. Target and the Environment (2013).
3. CorporateRegister.com (2013).
4. Procter & Gamble (2012).
5. Method (2013).
6. Tate, Ellram, and Dooley (2011).
7. Ball (2009).
8. Tate, Ellram, and Dooley (2011)
9. Green Logix (2013).
10. Ball (2009); Green Logix (2013).
11. Sustainability Consortium (2013).
12. Patagonia (2013).
13. Ball (2009); Patagonia (2013).
14. Packaging-Gateway.com (2012).

References

American Waterways (March 2009). *A modal comparison of domestic freight transportation effects on the general public.* Center for Ports and Waterways, Texas Transportation Institute. Accessed January 25, 2013 at http://www.americanwaterways.com/press_room/news_releases/NWFSTudy.pdf

Apple (2012). *Supplier responsibility report, Auditing.* Accessed January 10, 2013 at http://www.apple.com/uk/supplierresponsibility/auditing.html

Apple (2012). *Supplier responsibility report, Environmental impact.* Accessed January 10, 2013 at http://www.apple.com/uk/supplierresponsibility/code-of-conduct/environmental-impact.html

Ball, J. (2009). Six products, six carbon footprints. *Wall Street Journal.* Accessed March 20, 2013 at http://online.wsj.com/article/SB122304950601802565.html

Bilodeau, J. (2013). Developing technology strategies for reverse logistics. *Reverse Logistics Digital Magazine*, January 7. Accessed January 10, 2013 at http://www.rlmagazine.com/

Brown, B. (2011). *Researchers waste management pyramid takes aim at messy computer innards*, September 2. Accessed on January 24, 2013 at http://www.networkworld.com/community/blog/researchers'-waste-management-pyramid-takes-a

BSR (2012). *Business for social responsibility*, Accessed February 2, 2012 from http://www.bsr.org/

BSR (2013). *Clean Cargo Working Group.* Accessed January 24, 2013 at http://www.bsr.org/our-work/working-groups/clean-cargo

Business Dictionary (2012). *Waste reduction.* Accessed November 25, 2012 at http://www.businessdictionary.com/definition/waste-reduction.htmlon

Callahan, S. J., & Tohomas, J. M. (2009). Module 1: Modeling environmental problems. *Environmental Economics & Management: Theory, Policy, and Applications.* Cengage.

Carbon Neutral Company (2013). Accessed January 31, 2013 at http://www.carbonneutral.com/our-services/measure/

Carlson, J. (2013). *The sustainability measurement and reporting system (SMRS).* The Sustainability Consortium. Accessed January 22, 2013 at http://www.sustainabilityconsortium.org/smrs/

Carson, R. (1962). *Silent Spring,* Houghton Mifflin.

Carter, C. R., & Carter, J. R. (1998). Interorganizational determinants of environmental purchasing: Initial evidence from the consumer products industries. *Decision Sciences 29*(3), 659–684.

Certified B Corporation (2013). *Why B corps matter.* Accessed January 24, 2013 at http://www.bcorporation.net/what-are-b-corps/why-b-corps-matter

Cheeseman, G.M. (2010). *Is Walmart's sustainability consortium a genuine effort to develop better products?* Triple Pundit, February 9. Accessed January 22, 2013 at http://www.triplepundit.com/2010/02/is-walmarts-sustainability-consortium-a-genuine-effort-to-develop-better-products/

CHEP (2013). *About CHEP.* Accessed January 31, 2013 at http://www.chep.com/

Churchill, J. (2012). *Filling shipping's $1 billion hole—the logistical challenge of empty shipping containers,* March 12. Accessed February 2, 2013 at http://gcaptain.com/filling-shippings-billion-hole/

Clean Energy Ideas (2012). *Definition of recycling,* November 27. Accessed March 27, 2013 at http://www.clean-energy-ideas.com/energy_definitions/definition_of_recycling.html

Coca-Cola Company (2013). *Water Conservation.* Accessed April 2, 2013 at http://www.livepositively.com/en_us/water_conservation/#/water_conservation_overview

CorporateRegister.com (2013). *About corporate register.* Accessed March 20, 2013 at http://www.corporateregister.com/

Davis, S. C., Diegel, S. W., & Boundy, R. G. (2011). *Transportation energy data book.* U.S. Department of Energy, Washington, DC.

Dell (2006). *Dell sustainability report: Dell fiscal year 2006 in review.* Accessed May 27, 2008 at http://www.corporateregister.com/a10723/dell06-sus-usa.pdf

Dell (2012). *Dell highlights environmental, giving and community work in FY12 corporate responsibility report,* July 18. Accessed January 17, 2013 at http://content.dell.com/us/en/corp/d/secure/2012-07-18-dell-corporate-responsibility-report

Duke, M. (2012). *New commitments to drive sustainability deeper into Walmart's global supply chain,* October 12. Accessed January 22, 2013 at http://news.walmart.com/executive-viewpoints/new-commitments-to-drive-sustainability-deeper-into-walmarts-global-supply-chain

Dutch Waste Management Association (2012). *Waste matters.* Accessed November 25, 2012 at http://www.wastematters.eu/news-from-europe/news-from-europe/uneven-playing-field-for-landfill-in-europe.html

EICC (2012). *List of members.* Accessed February 1, 2013 at http://www.eicc.info/documents/EICCMembers08-31-12.pdf

EICC (2012). *Code of conduct,* November 12. Accessed January 10, 2013 at http://www.eicc.info/documents/CodeofConduct30vs40FINAL.pdf

Ellram, L. M., & Golicic, S. (2013). *Adopting environmentally sustainable transportation practices: The role of voluntary partnerships.* Working paper.

Ellram, L. M., & Tate, W. (2003). IBM's supply-base efforts towards sustainability. *PRACTIX: Best Practices in Purchasing and Supply Management, 6*(4).

Engel, B. (2011). 10 Best practices you should be doing now. November 25, 2012. *Supply Chain Quarterly*. Accessed at http://www.supplychainquarterly.com/topics/Procurement/scq201101bestpractices/

Environmental Agency (2013). *Waste Crime Intervention Program*. End of Life Vehicle Information for authorized treatment facilities. Accessed January 14, 2013 at http://a0768b4a8a31e106d8b0-50dc802554eb38a24458b98ff72d550b.r19.cf3.rackcdn.com/geho0411btwg-e-e.pdf

Environment Agency Europe (2012). *Environmental permitting regulations*, November 27. Accessed at http://scp.eionet.europa.eu/definitions/recovery

Environmental Leader (2008). *IBM, GSK, Herman Miller see healthy ROI from green purchasing*, May 8. Accessed January 10, 2013 at http://www.environmentalleader.com/2008/05/08/ibm-gsk-herman-miller-see-healthy-roi-from-green-purchasing/

EPA (2012). *Sustainability*. Accessed October 13, 2012 at http://epa.gov/sustainability/basicinfo.htm

EPA (2012). *Environmentally preferable purchasing, electronic product environmental assessment Tool*, October 12. Accessed January 17, 2013 at http://www.epa.gov/epeat

EPA (2012). *EPA's greenhouse gas emission reductions*, November 12. Accessed January 15, 2013 at http://www.epa.gov/oaintrnt/ghg/index.htm

EPA (2012). *Air and radiation: Basic information*, United States environmental protection agency, November 30. Accessed April 2, 2013 at http://www.epa.gov/air/basic.html#aqi

EPA (2013). *Risk management, sustainable technology: Lifecycle perspective*. Accessed January 31, 2013 at http://www.epa.gov/nrmrl/std/lifecycle.html

EPA (2013). *Effects of air pollutants—health effects*. Accessed January 25, 2013 at http://www.epa.gov/eogapti1/course422/ap7a.html

EPA (2013). *Environmentally preferable purchasing*. Accessed January 17, 2013 at http://www.epa.gov/epp/

EPA (2013). *Statistics on the management of used and end-of life electronics*. Accessed January 17, 2013 at http://www.epa.gov/osw/conserve/materials/ecycling/manage.htm

EPA (2013). *Hazardous Materials Transportation Act*. Accessed January 25, 2013 at http://www.epa.gov/osweroe1/content/lawsregs/hmtaover.htm

EPA (2013). *SmartWay*. Accessed January 25, 2013 at http://www.epa.gov/smartway/

EPA (2013). *SmartWay*, SmartWay transport partnership, carrier performance rankings. Accessed January 25, 2013 at http://www.epa.gov/smartway/partnership/performance.htm

EPEAT (2013). *Environmental criteria.* Accessed January 17, 2013 at http://www.epeat.net/resources/criteria-discussion/

EPEAT (2013). *Verification and criteria.* Accessed January 28, 2013 at http://www.epeat.net/resources/criteria-verification/

E-stewards (2013). The e-waste crisis. *Basil Action Network.* Accessed January 17, 2013 at http://e-stewards.org/the-e-waste-crisis/

European Union (2012). *Product Environmental Footprint.* Accessed February 4, 2013 at http://ec.europa.eu/environment/eussd/product_footprint.htm

European Union (2013). *Waste and electronic equipment.* Accessed January 17, 2013 at http://ec.europa.eu/environment/waste/weee/legis_en.htm

Eurostat (July 2011). *Waste Statistics.* Accessed January 31, 2013 at http://epp.eurostat.ec.europa.eu/statistics_explained/index.php/Waste_statistics

Gerrard, J., & Kandlikar, M (2007). Is European end-of-life vehicle legislation living up to expectations? Assessing the impact of the ELV directive on 'green' innovation and vehicle recovery. *Journal of Cleaner Production* 15, 17–27.

Green Logix (2013). *Fleets investing in green truck driver training will reduce fuel costs.* Accessed February 4, 2013 at http://www.fieldtechnologies.com/fleets-investing-in-green-truck-driver-training-will-reduce-fuel-costs/

Hewlett-Packard (2006). *Hewlett-Packard 2006 citizenship report web content.* Accessed May 27, 2008 at http://www.corporateregister.com/a10723/Hewlettpack06-sus-usa.pdf

Hewlett-Packard (2011). *2011 Global citizen report: Environmental sustainability.* Accessed January 17, 2013 at http://www.hp.com/hpinfo/globalcitizenship/society/supplychain.html

Hewlett-Packard (2011). 2011 *Global citizen report: Product reuse and recycling.* Accessed January 17, 2013 at http://www.hp.com/hpinfo/globalcitizenship/media/files/hp_fy11_gcr_product_reuse_and_recycling.pdf#PS9.indd:PS9

Hewlett-Packard (2012). *Product design for environment.* HP Environment. Accessed November 30, 2012 at http://www.hp.com/canada/corporate/hp_info/environment/products/design-for-environment.html

Himanen, V., Lee-Gosselin, M., & Perrels, A (2004). Impacts of transport on sustainability: towards an integrated transatlantic evidence base. *Transport Reviews* 24(6), 691–705.

Honda (2012). *Reporting— Honda-scope 3.* Accessed January 17, 2013 at http://world.honda.com/news/2012/c120825Greenhouse-Gas-Emissions/index.html

Honda (2013). *Honda product recycling (3Rs).* Accessed January 14, 2013 at http://world.honda.com/environment/activities/product_recycling.html

IBM (2002). *IBM environmental requirements for materials, parts, and products.* IBM Corporation. Accessed March 20, 2013 at http://www.ibm.com/ibm/environment/products/ecpquest.shtml

IBM (2011). *Corporate responsibility report; supply chain—supplier assessment and improvement plans.* Accessed January 10, 2013 at http://www.ibm.com/ibm/responsibility/2011/supply-chain/supplier-assessment-and-improvement-plans.html

IBM (2013). *Position on scope 3 GHG emissions.* Accessed January 31, 2013 at http://www.ibm.com/ibm/environment/climate/scope3.shtml

International Energy Agency (2008). *Tracking industrial energy efficiency and CO_2 emissions.* (Paris, France: OECD/IEA).

Iobst, E. (2012). A small company's journey to sustainability. *Supply Chain Management*, October, Executive Speaker Series, Miami University, Farmer School of Business.

Jackson, N. (2011). A conversation with Kevin Dooley, sustainability consortium director. *Atlantic Monthly*, November 11. Accessed January 10, 2013 at http://www.theatlantic.com/national/archive/2011/11/a-conversation-with-kevin-dooley-sustainability-consortium-director/247655/

Joyner, A. (2012.). Coyote Logistics zooms toward $1 billion, June 11. Accessed January 26, 2013 at http://www.inc.com/best-industries-2012/april-joyner/coyote-logistics-zooming-toward-1-billion.html

Just Live Greener (2012). *Scary facts about landfills*, November 25. Accessed March 25, 2013 at http://justlivegreener.com/environment/253-scary-facts-about-landfills.html

Kanari, N., Pineau, J. L., & Shallari, S. (2003). Overview: Recycling end-of-life vehicle recycling in the European Union. *JOM*, August (55:8). Accessed March 21, 2013 at http://www.tms.org/pubs/journals/jom/0308/kanari-0308.html

Kraft Foods (2012). *Kraft Foods waste reduction sustainability success stories.* Accessed January 31, 2013 at http://www.kraftfoodscompany.com/SiteCollectionDocuments/corp/waste_reductions_factsheet.pdf

Kraljic, P. (1983). Purchasing must become supply management. *Harvard Business Review*, *61*(5), 109–117.

Krause, D. R., Vachon, S., & Klassen, R. D. (2009). Special topic forum on sustainable supply chain management: introduction and reflections on the role of purchasing management. *Journal of Supply Chain Management 45*(4), 18–25.

Leavoy, P. (2011). *Is ISO 14001 certification still relevant?* Greenbiz.com, August 4. Accessed January 10, 2013 at http://www.greenbiz.com/blog/2011/08/04/iso-14001-certification-still-relevant

LeBlanc, R. (2012). *Honda logistics offers customized 3PL services.* Packaging Revolution. Accessed January 21, 2012 at http://packagingrevolution.net/honda-logistics-offers-customized-3pl-services-and-patented-reusable-crates-to-mid-sized-companies/

Lee, H, O'Marah, K., & John, G. (2012). The chief supply chain officer report 2012. *SCM World.* September 2012, p. 29.

Lexus (2013). *Lexus environment protection.* Accessed January 14, 2013 at http://www.lexus.com.cn/en/brand/zzjy.html

Manners Bell, J. (2010). *Move to ocean: Dell's massive shift in supply chain strategy, July 10.* Transport Intelligence. Accessed July 18, 2010 at http://www.transportintelligence.com/briefs/move-to-ocean-dells-massive-shift-in-supply-chain-strategy/1915/

McDonough, W., & Braungart, M. (2002). *Cradle to Cradle,* Northpoint Press.

Mersen (2013). *Does RoHS apply to you?* Accessed January 17, 2013 at http://ep-us.mersen.com/compliance/rohs-apply.html

Method (2013). *Biodiesel fleet.* Accessed January 26, 2013 at http://methodhome.com/greenskeeping/biodiesel-fleet/

Method (2013). *C2C process.* Accessed at http://methodhome.com/greenskeeping/c2c-process/

Method (2013). *Defining sustainability.* Accessed January 24, 2013 at http://methodhome.com/greenskeeping/defining-sustainability/

Method (2013). *Greenskeeping what we do—packaging.* Accessed January 29, 2013 at http://methodhome.com/greenskeeping/our-packaging/

Method (2013). *Our distribution.* Accessed January 26, 2013 at http://methodhome.com/greenskeeping/our-distribution/

Method (2013). *Our incentives.* Accessed January 24, 2013 at http://methodhome.com/greenskeeping/our-incentives/

Motorola (2013). *Motorola and the environment.* Accessed February 4, 2013 at http://responsibility.motorola.com/index.php/environment/operations/

MSNBC (2013). *Is Wal-Mart going green.* MSNBC.com News. Accessed January 21, 2013 at http://www.msnbc.msn.com/id/9815727/ns/us_news-environment/t/wal-mart-going-green/

Nandagopal, R., & Sankar, A. (2009). Wal-Mart's environmental strategy. *Asian Journal of Management Cases 6,* 119. Accessed January 21, 2013 at http://www.cfsgs.com/uploads/3/6/0/1/3601225/119.pdf

NASPO (2013). *Why is it important to buy green.* Accessed January 31, 2013 at http://www.naspo.org/content.cfm/id/Green_Guide_Basics

Noharm.org (2013). *Issues: green purchasing.* Adapted from Healthcare Without Harm. Accessed January 31, 2013 at http://www.noharm.org/europe/issues/purchasing/guide.php

Narasimhan, R., & Carter, J.R. (1998). *Environmental Supply Chain Management.* Center for Advanced Purchasing Studies, Tempe, AZ.

Nokia (2013). *About Nokia.* Accessed February 1, 2013 at http://www.nokia.com/global/about-nokia/people-and-planet/strategy/casestudies/supplier-environmental-audits/

NRDC (2012). *Clean by design: transportation. Developed from the following sources by Natural Resources Defense Council: 3.* Efficiency figures for Truck and Air from: EIA. Annual Energy Outlook. 2011, tables 66, 68. Efficiency figure for Rail from: DOE, "Transportation Energy Data Book," Ed. 29, 2010, table 2.16. Efficiency figure for large container ship calculated from EMEP/EEA, "Emission Inventory Guidebook 2009," updated Mar 2011. Emissions factors for Truck, Rail, and Air for CO2, N2O, CH4 from: EIA, "Documentation for Emissions of Greenhouse Gasses in the US 2005." PM10 emissions factors for Air and Truck adapted from Facanha & Horvath, "Evaluation of Life-Cycle Air Emission Factors of Freight Transportation," *Environmental Science Technology*, 2007, 41, 7138–7144. PM10 emission factor for Rail from EPA, Technical Highlights: Emissions Factors for Loco-motives, 2009, Table 6. Emission factors for ship from: IMO, "Second IMO GHG Study 2009," Table 3.6. Upstream fuel and feedstock emissions from: Argonne Labs, GREET 1.8 model. Accessed January 25, 2013 from Natural Resources Defence Council at http://www.nrdc.org/international/cleanbydesign/transportation.asp

NTL (April 2010). Accessed January 25, 2013 at http://ntl.bts.gov/lib/32000/32700/32779/DOT_Climate_Change_Report_-_April_2010_-_Volume_1_and_2.pdf

O'Reilley, J. (June 2012). *Green is still white hot. Inbound Logistics.* Accessed January 25, 2013 at http://www.inboundlogistics.com/cms/article/green-is-still-white-hot/

Packaging-Gateway.com (March 2012). Material world: innovative packaging components. Accessed February 4, 2013 at http://www.packaging-gateway.com/features/featurematerial-world-innovative-packaging-components

Pagell, M., Wu, Z., & Wasserman, M.E. (2010). Thinking differently about purchasing portfolios: an assessment of sustainable sourcing. *Journal of Supply Chain Management 46*(1), 57–73.

Patagonia (2013). *Changing ports pays dividends.* Accessed February 4, 2013 at http://www.patagonia.com/us/patagonia.go?assetid=79365

Patagonia. (2013) *The footprint chronicles.* Accessed February 4, 2013 at http://www.patagonia.com/us/footprint/

PCF World Forum (2012). *Product environmental footprint FAQ.* Accessed at http://www.pcf-world-forum.org/about/product-environmental-footprint-faq/ and Product Environmental Footprint.

Pelley, S. (2009). *The wasteland.* 60 Minutes, August 30. Accessed March 20, 2013 at http://www.cbsnews.com/video/watch/?id=5274959n

Plambeck, E., & Denend, L. (2007). *Walmart's sustainability strategy* (A). Stanford University, GSB No. OIT-71A, February. Accessed January 21, 2013 at http://cb.hbsp.harvard.edu/cb/web/product_detail

Procter & Gamble (2012). *P&G sustainability overview.* Accessed January 23, 2013 at http://www.pg.com/en_US/sustainability/overview.shtml

Procter & Gamble (2013). *Detergent compaction: Why should we be concerned about detergent compaction?* Accessed March 20, 2013 at http://scienceinthebox.com/concentrated-detergents

Procter & Gamble (2013). *Sustainable packaging: Why should we be concerned about sustainable packaging?* Accessed March 20, 2013 at http://scienceinthebox.com/environmentally-friendly-packaging

Resourcities (2012). *Resourcities.* Accessed October 10, 2012 at http://resourcities.acrplus.org/waste_resources/europe_waste.htm

Rocky Mountain Institute (2013). *Trucking.* Accessed January 25, 2013 at http://www.rmi.org/Trucking

Rocky Mountain Institute (2013). *Rocky Mountain Institute vision and mission.* Accessed January 25, 2013 at http://www.rmi.org/Vision%20and%20Mission

Rogers, D.R. & Tibben-Lembke, R. (1999). Going backwards: reverse logistics trends and practices. Reverse Logistics Executive Council.

Rosenbloom, S. (2010). Wal-Mart unveils plan to make supply chain greener, February 25. *New York Times.* Accessed January 21, 2013 at http://www.nytimes.com/2010/02/26/business/energy-environment/26walmart.html?_r=0

ShippingEfficiency.org (2013). *Savings calculator.* Accessed January 29, 2013 at http://shippingefficiency.org/savings-calculator

Smock, D. (2007). Efforts grow to design for disassembly. *Design News*, November 19. Accessed January 17, 2013 at http://www.designnews.com/document.asp?doc_id=213642

Sprint (2013). *Re:cycle.* Accessed January 31, 2013 at http://www.sprint.com/responsibility/communities_across/index.html?ECID=vanity:recycle

Starbucks (2012). *Save money, cups and the planet*, November 25. Accessed at http://blogs.starbucks.com/blogs/customer/archive/2012/10/02/save-money-cups-and-the-planet.aspx

Straight, B. (2010). Wal-Mart seeking greater freight control. *Fleet Owner.* Accessed January 28, 2013 at http://fleetowner.com/management/news/walmart-greater-freight-control-0611

Sunny Delight (2011). Improving our footprint step by step. *Sustainability Report.* Accessed March 25, 2013 at http://ww2.sunnyd.com/files/2011_sustainability_report.pdf November 30, 2012. Sustainability Consortium (2013). Open IO Beta 1.5. Accessed February 4, 2013. http://www.sustainabilityconsortium.org/open-io/

Sustainability Consortium (2013). *Use the model.* Accessed January 22, 2013 at http://www.sustainabilityconsortium.org/open-io/use-the-model/

Target (2013). *Target and the environment.* Accessed February 4, 2013 at https://corporate.target.com/corporate-responsibility/environment

Tate, W. L., Dooley, K. L., & Ellram, L. M. (2011). Drivers of supplier adoption of sustainable business practices, *Journal of Business Logistics*, *32*(1), 6–16.

Tate, W. L., Ellram, L. M., & Dooley, K.J. (2012). Environmental purchasing and supplier management (EPSM): Theory and practice. *Journal of Purchasing and Supply Management*, *18*(3), 173–188.

Texas Transportation Institute (2011). *Inland waterways working for America*, barge emissions adapted as ratio of truck emissions. Accessed March 20, 2013 at http://www.marad.dot.gov/documents/water_works_REV.pdf

U.K. National Measurement Office (2013). *RoHS compliance*. Accessed January 17, 2013 at http://www.rohscompliancedefinition.com/

UNEP/ SETAC Life Cycle Initiative (2013). *Life cycle approaches: the road from analysis to practice*. p. 14. Accessed January 31, 2013 at http://www.unep.fr/shared/publications/pdf/DTIx0594xPA-Road.pdf

U.S. Census Bureau (2010). *Survey of Manufactures*. Accessed January 30, 2013, http://www.census.gov/manufacturing/asm/

U.S. Census Bureau (2012). *Latest annual retail trade report*, March 12. Accessed January 30, 2013 at http://www.census.gov/retail/index.html#

USCar.org (2013). *Vehicle recycling partnership video*. Accessed January 14, 2013 at http://www.uscar.org/guest/article_view.php?articles_id=233

USCar.org (2013). *Vehicle recycling partnership*. Accessed January 14, 2013 at http://www.uscar.org/guest/teams/16/Vehicle-Recycling-Partnership

Walmart (2007). *Concentrated liquid laundry detergent*. Accessed January 23, 2013 http://www.walmartfacts.com/Media/128352917794036250.pdf

Walmart (2008). *Walmart announces global responsible sourcing initiative at China summit*, October 22. Accessed January 30, 2013 at http://www.csrwire.com/press_releases/24229-Wal-Mart-Announces-Global-Responsible-Sourcing-Initiative-at-China-Summit-

Walmart (2009). *Walmart announces sustainable product index*. Accessed January 22, 2013 at http://news.walmart.com/news-archive/2009/07/16/walmart-announces-sustainable-product-index

Walmart (2010). *Walmart announces goal to eliminate 20 million metric tons of greenhouse gas emissions from global supply chain*, February 25. Accessed January 21, 2013 at http://news.walmart.com/news-archive/2010/02/25/walmart-announces-goal-to-eliminate-20-million-metric-tons-of-greenhouse-gas-emissions-from-global-supply-chain

Walmart (2012). *Walmart's 2012 global responsibility report*. Accessed November 30, 2012 at http://corporate.walmart.com/global-responsibility/environment-sustainability/global-responsibility-report

Walmart (2013). *Truck fleet*. Accessed January 28, 2013 at http://corporate.walmart.com/global-responsibility/environment-sustainability/truck-fleet

Wasik, J. F. (2010). *The surprising success of the green supply chain*, September 9. Accessed January 15, 2013 at http://money.cnn.com/2010/08/13/news/companies/corporate_sustainability.fortune/index.htm

Wilson, R. (2012). *23rd Annual State of Logistics Report*, June 18. Presented to the National Press Club, Washington DC. Accessed June 18, 2012 at www.cscmp.org

Wu, H., & Dunn, S. (1995). Environmentally responsible logistics systems. *International Journal of Physical Distribution and Logistics Management 25*(2), 20–38.

Wolf, C., Burritt, C., & Boyle, M. (2010), Why Wal-Mart wants to take the driver's seat, May 27. *BusinessWeek*. Accessed January 28, 2013 at http://www.businessweek.com/magazine/content/10_23/b4181017589330.htm

World Bank (2013). Accessed February 2, 2013 at http://siteresources.worldbank.org/INTAIRTRANSPORT/Resources/TP38.pdf

World Shipping Council (2011). *2011 container supply review*. Accessed February 2, 2013 at http://www.worldshipping.org/public-statements/2011_Container_Supply_Review_Final.pdf

Zsidisin, G. A., & Siferd, S. P. (2001). Environmental purchasing: A framework for theory development. *European Journal of Purchasing & Supply Management 7*(1), 61–73.

Index

OTHER TITLES IN THE SUPPLY
AND OPERATIONS MANAGEMENT COLLECTION

Johnny Rungtusanatham, The Ohio State University, Collection Editor

- *Transforming U.S. Army Supply Chains: Strategies for Management Innovation* by Greg Parlier
- *Design, Analysis and Optimization of Supply Chains: A System Dynamics Approach* by William Killingsworth
- *Supply Chain Planning and Analytics: The Right Product in the Right Place at the Right Time* by Gerald Feigin
- *Supply-Chain Survival in the Age of Globalization* by James A. Pope
- *Better Business Decisions Using Cost Modeling: For Procurement, Operations, and Supply Chain Professionals* by Victor Sower and Christopher Sower
- *Supply Chain Risk Management: Tools for Analysis* by David L. Olson
- *Leading and Managing the Lean Management Process* by Gene Fliedner
- *Supply Chain Information Technology* by David L. Olson
- *Global Supply Chain Management* by Matt Drake
- *Managing Commodity Price Risk: A Supply Chain Perspective* by George A. Zsidisin and Janet L. Hartley
- *Improving Business Performance with Lean* by James Bradley
- *RFID for the Supply Chain and Operations Professional* by Pamela Zelbst and Victor Sower
- *Insightful Quality: Beyond Continuous Improvement* by Victor Sower and Frank Fair
- *Sustainability Delivered: Designing Socially and Environmentally Responsible Supply Chains* by Madeleine Pullman and Margaret Sauter
- *Strategic Leadership of Portfolio and Project Management* by Timothy J. Kloppenborg and Laurence J. Laning
- *Sustainable Operations and Closed-Loop Supply Chains* by Gilvan C. Souza
- *Mapping Workflows and Managing Knowledge: Capturing Formal and Tacit Knowledge to Improve Performance* by John L. Kmetz
- *Supply Chain Planning: Practical Frameworks for Superior Performance* by Matthew Liberatore and Tan Miller
- *Understanding the Dynamics of the Value Chain* by William Presutti and John Mawhinney
- *An Introduction to Supply Chain Management: A Global Supply Chain Support Perspective* by Edmund Prater and Kim Whitehead
- *Project Strategy and Strategic Portfolio Management: A Primer* by William H. A. Johnson and Diane Parente
- *Designing Supply Chains for New Product Development* by Antonio Arreola-Risa and Barry Keys
- *Production Line Efficiency: A Comprehensive Guide for Managers* by Sabry Shaaban and Sarah Hudson

Announcing the Business Expert Press Digital Library

Concise E-books Business Students Need
for Classroom and Research

This book can also be purchased in an e-book collection by your library as

- a one-time purchase,
- that is owned forever,
- allows for simultaneous readers,
- has no restrictions on printing, and
- can be downloaded as PDFs from within the library community.

Our digital library collections are a great solution to beat the rising cost of textbooks. e-books can be loaded into their course management systems or onto student's e-book readers.

The **Business Expert Press** digital libraries are very affordable, with no obligation to buy in future years. For more information, please visit **www.businessexpertpress.com/librarians**. To set up a trial in the United States, please contact **Adam Chesler** at *adam.chesler@businessexpertpress.com* for all other regions, contact **Nicole Lee** at *nicole.lee@igroupnet.com*.

www.ingramcontent.com/pod-product-compliance
Lightning Source LLC
Chambersburg PA
CBHW071200200326
41519CB00018B/5294